Moving

Through

Parallel Worlds

To

Achieve

Your

Dreams

The Epic Guide To Unlimited Power

Kevin L. Michel

ACKNOWLEDGEMENTS

Thank you to you the reader, for having the courage to purchase a book that is beyond the boundary of conventional thinking and approaches. A book that is written plainly so that it can reach many, but most certainly not plainly enough. This book is at times very technical to highlight the academic basis for many of the ideas discussed. This book is dedicated to you the reader, and to the memory of the great physicist Hugh Everett III.

I must thank the many authors who I quote throughout this book and thank also the many authors mentioned as sources. Our approach to the topic of Moving Through Parallel Worlds to Achieve Your Dreams is both about a new way of thinking, that is, a new paradigm and also about a very scientific and academic approach to a metaphysical topic. Because of this academic approach I have used quotations and references from other authors to give the reader an understanding of the concepts at play. So this work is academic and also metaphysical and theoretical. It really is not possible to communicate the complex scientific ideas discussed in this book without deferring to the research, ideas and quotations of the leading thinkers in fields ranging from quantum physics and psychology to biology and epigenetics. I thank the numerous other authors for the material provided in fair-use and have acknowledged their work throughout the text and as sources in the reference section where applicable. My inclusion of quotes, or references to a particular scientist's experiment or study, is not meant to suggest an endorsement of my ideas by the mentioned author or scientist.

I make a particular mention of the legendary scientist Hugh Everett III, who wrote a breathtaking and mathematically rigorous paper introducing the physics community to what subsequently became called: 'The Many Worlds Interpretation of Quantum Mechanics.' Hugh Everett III lived from November 11th 1930 to July 19th 1982. My references to Dr. Everett's work do not presuppose his endorsement of this book, but conversely, this book is my endorsement and celebration of the spirit and energy of a brilliant physicist. Hugh Everett III was willing to go against the grain and willing to stand up for an idea that emerged from his analysis of the facts, and did so in spite of resistance from the status quo. May he be well remembered.

CONTENTS

FOREWORD

Who among us has had every experience in human existence? Which one of us has had such astounding success in life that we have garnered the privilege to bestow wisdom upon the masses? This work has emerged from all that I have learnt from great teachers throughout human history. Human beings who were greatly wise and who saw much. This work has emerged from my own awareness and realizations. This work has emerged from my stumbling onto a connection between my mind and all minds; from a spiritual connection through unexpected insights and a real physical connection through research into the lives, stories and insights of great achievers, scientists, and spiritual teachers. This work has emerged from my focus on the lessons from all the little that I have seen. Garnering from the smallest of experiences, the greatest of spiritual lessons, and physical lessons. To see in the simplest of scientific experiments, broader implications. To see the connection between the quantum world and the macroscopic world of the human senses. To see the future.

The concepts in this book came through me like a rushing torrent, out of control, free-flowing and all consuming, in an intense, solitary period of insight. After a twelve hour session of typing I would read back the text, and notice that I would consistently refer to myself as 'We.' At first, I would diligently change all the 'we' pronouns to 'I's but then, I realized. I was being used by the work, more than I was creating it. Both the science and the metaphysics had something to say, something to share, something to communicate and I became the curious conduit.

This book is for you. It is for those of you who work on the front lines. To clerks and future change agents, and those who would be sent off to fight the wars. To the oft overlooked and ignored, and seemingly insignificant young dreamers. For you, the graduate. This book is for those lost or just delayed on an uncommon or unexpected path. This book is for those who still believe in their greatness but need help finding the way. This is your guide to transformation, transcendence and power. This is the starting point of your mastery of wealth, creation, and mastery of all things in the physical world. This book is all about bridging the discontinuity in your life from the point where you are at now to the point where you dream that you can be. This book is about putting you into alignment with all that you have imagined possible for yourself and showing you a path even to that which you may have considered impossible. This book is so you may be lifted up and realize the power you have to exist in a world that is as you imagine it should be. The tools and paradigms in this book will allow you to achieve greatly and alter the trajectory of your life as you continue your existence through parallel worlds. Let the adventure begin!

STAGE 1

Starting Your Adventure Through Parallel Worlds

"The biggest adventure you can take is to live the life of your dreams." - Oprah Winfrey

THIS BOOK

This book is science and it is also meditation. It is a guidebook and it is an extended mantra. This book is academic and it is metaphysical. This book is a scientific, meditative, academic, metaphysical adventure. This work will allow you to make the connections in your own mind that will significantly strengthen your ability to achieve any desired ambition. This book is the beginning of an adventure. An adventure that will branch across many worlds, with my promise that you will in the process, hasten your own personal evolution and accentuate your power.

This book will allow you to move through parallel worlds whilst retaining the elements of the various worlds that are most pleasing to you. In very quick time, this book will allow you to connect with the untapped powers of your mind, the untapped powers of your soul and with the very essence of your being as an all-powerful and divine creator.

I want to begin by promising you that if you seek to understand this book, and to learn from it and you apply it, I promise that you will come away with the key to moving between parallel worlds to arrive at a place which is the Ideal Parallel World of your choosing. Do not be concerned if some of the science appears too complicated the first time you are exposed to it. Indeed, much of the information, particularly the quantum physics elements, go against most people's current view of the world. Apropos, if these ideas were easy to grasp then everyone else would have already embraced them, which would negate the advantage gained from reading this now.

WHAT IS A PARALLEL WORLD?

A parallel world is one of an infinite series of non-interacting physical structures that represents the totality of an individual's consciousness and existence. 'Parallel world' refers to what is often called a 'parallel universe,' but we use the term 'world' and suggest that the totality of all worlds would be what makes up your single universe.

A world is a system that connects you to one of many points of quantum 'superposition'[D1] that a quantum particle exists in at the point of human awareness and subsequent 'collapse of the wave function.'[D2] Let us define two of these terms/phrases, because they will occur throughout the book, and then we shall also look at our 'parallel world' definition in less technical language.

D1: Superposition : soo-pər-pə-zi-SHən

1. The existence of a single particle/object/world in more than one location simultaneously.

- 'Superposition' occurs with quantum sized particles in the 'double-slit experiment.'

2. The existence of a single particle/object/world in more than one state simultaneously, as in a wave existence and a particle existence.

- For example, if you are sitting in your bedroom and reading this at your desk, whilst at that exact moment, you are also sitting in your dream house fifty miles away reading this – then you are in superposition – two or more places at once.

D2: Collapse of the wave function

1. This is the process where a particle/object/world shifts from a wave of possible locations to just one location. This is called 'decoherence' in quantum physics.

2. The shift of one's reality from an infinity of possibilities and probabilities to the single reality that is observed.

3. The shift from superposition, to a single position.

Continuing to define 'parallel worlds;' the parallel worlds that one moves through are worlds that exist in a real physical sense as described in the 'Many-Worlds Interpretation of Quantum Mechanics,' and/or worlds that exist only in the mind of the individual in the 'Many-Minds Interpretation of Quantum Mechanics.' The parallel worlds we describe are theoretically testable, although a suitable test has not yet been devised by quantum physicists. However, this current lack of an adequate test is fundamentally irrelevant to the ambition of the individual looking to achieve their dreams by moving through parallel worlds.

In that, if you are caught up in an uninspiring job, on a mediocre income - or if you are a recent high-school or college graduate just getting started but not feeling like you are on the right path, and through reading and applying the concepts in this book you start earning millions annually, then have a dream house and dream spouse and a dream career, then I suspect you would not bother to have a debate on semantics, testability, and definitions. In that, you would not bother to define whether or not you are in a parallel world, or whether you are just doing very well in the same world where you started. The idea of parallel worlds is that you are living one ordinary existence now and through the application of the ideas in this book, you shift from the plane of that ordinary existence to a life that is aligned with your dreams. That, just described, is a movement through parallel worlds. Although, as we shall see, this parallel world movement is also meant in a literal sense.

THE LANGUAGE OF THIS BOOK

There will be numerous times in this book where I shall describe a process in a manner that is metaphysical, that is, spiritual or otherworldly, by speaking of the spirit, the source, God or god, infinite intelligence, the matrix of the universe, or innumerable other terms that may or may not connect with you directly. Some readers will be able to go right along with a metaphysical concept because they feel that it speaks to them directly, and some may claim an agnosticism or atheism towards such beliefs. This is entirely acceptable and will not reduce the effectiveness of this book in providing you with the results and achievements that you seek. If a spiritual term does not speak to you directly, then I ask you to view that term as a metaphor and not as literal.

I have no memory of life before earth and no knowledge of that which occurs after; I can only speak from the impressions that I have received as human. If I say, you 'must keep your spirit strong,' and you do not believe in the concept of 'spirit,' then you can think of your 'spirit' as your energy, or your core, or your subconscious mind or essentially whatever works for you. I have found that non-believers in faith often can draw from spiritual teachings even when they do not share a literal belief in the specific theology. If you were dehydrated in the desert and a man came along and handed you a glass of water, but he called the water by some odd name, you need not correct him - the effect on your wellness, remains the same.

Conversely, there will be scientific references in this book that believers of some faiths may not agree with. If I state that the geological reference indicates that the earth is more than 4 billion years old and your teaching suggests a much younger version of the earth – then integrate your view of the age of the earth into the idea, ignoring the detail but capturing the bigger point.

This book may in many ways exhaust readers of all faiths and readers of no faith; it will do so because it will force you to think both on specific details and ideas but also because its very style is designed to engage not just your conscious mind but also your subconscious mind. As a kid, you remember staring at those autostereogram pictures - viewing the surface image until the bigger picture in the background jumped up at you. This book will ultimately jump up at you, and you will then see the bigger picture, and you will see that part of your decision to enter the human condition is that you deliberately blind yourself from the truth of this world. And in that truth lies your power to transcend the physical and achieve a life that is on par with your spiritual greatness (this very sentence is a great practice exercise on choosing how to interpret metaphysical/spiritual words in a secular manner, if necessary).

This book will show you a truth and that truth will be different for each individual – the picture will be different for each individual, but the potential for achievement will be great for all.

Like hypnosis, this book only works if you want it to. You must decide what to take in, but when taken in holistically, it will put you on a powerful path and serve your evolution. Through this evolution of self you become more adept as a creator and are able to powerfully manifest a world of your imagination - you are able to get on a path of existence that is in line with what we will call your Ideal Parallel World (IPW).

This book will explain how you can seize control of your life and then alter the trajectory of your life. First I will get you through the rigorous exercise of reading the science, whilst I also explain why the science is important. Then we will explore the two powerful pillars of parallel world travel:

I. Subconscious Power

II. IPW Thinking

These two pillars will allow you to take control of your journey through parallel worlds. We shall end with a look at your 'Timeline for Success.'

I must again, strongly advise, that you read boldly through the science with the confidence that even though the details may not all be clear at first reading, that your constant exposure to these definitions and topics throughout the book will allow your mind to pull it all together in the end. It is only in the end that you will come away with the complete understanding necessary to achieve all that you can imagine. Read the entire book - or read nothing at all. Read big, or go home. Relax, as you read through the more rigorous scientific experiments, concepts and theories that are throughout the book and that we jump into right now. Enjoy the journey!

"A little learning is a dang'rous thing;
Drink deep, or taste not the Pierian spring:
There shallow draughts intoxicate the brain,
And drinking largely sobers us again." - Alexander Pope

Activity 1:

1. Close your eyes.

2. Consciously take control of your breathing.

3. Take several deep breaths and very consciously focus on each breath.

4. Remind yourself that you are reading this book to grow, learn, evolve, become more powerful, be inspired and to achieve your dreams.

5. Decide that you will read this book, and that you will find in it information that was created specifically for you. Information that will allow you to achieve every one of your objectives.

6. Ask yourself: Can the observer affect that which is observed? Can two people read the same book and come away with two very different meanings? You answered yes - with conditions, so then consider this: within these pages there is an idea - that idea requires you to read the entire work in order for it to be created. Consider, that you must intend, now – you must decide now, that you will retrieve this idea – this inspiration. If you fail to create an intention before you begin, then you risk not deriving the deeper meaning from this work.

If a single book can have different meanings based on who reads it, then that book is in fact created once on the page and then created again, albeit differently, in the mind of each reader. You are a co-creator of this book - I thank you, I ask you to decide in advance what you intend to create and I acknowledge that your ability to read with your eyes closed is very impressive.

You, as a co-creator, feel free to send me an email at: KevinLMichel@gmail.com. I appreciate any thoughts, stories, critiques and insights you would like to share and may well integrate those into updated versions of this book.

STAGE 2

Quantum Mechanics and the Science of Parallel Worlds

"The mind, once stretched by a new idea, never returns to its original dimensions."

— Ralph Waldo Emerson

Quantum Physics studies the smallest constituent parts of matter. To understand how we make the movements through parallel worlds, we must first view a Quantum Physics experiment called the 'Double-Slit Experiment' and then we shall review the 'Many-Worlds Interpretation of Quantum Physics.'

THE DOUBLE-SLIT EXPERIMENT

You shall understand the mechanism of the double-slit experiment, but its conclusions shall expand the dimensions of your mind and test your credulity. The stretching of the mind that occurs is the exact reason why I present the experiment - the result shattered my assumptions about the world and may shatter yours as well. To set the ground work, we begin with a basic version of the experiment performed by Thomas Young in the 1800s.

Thomas Young showed the result when we:

1. Shine a constant beam of light through a thin 'double-slit' opening in a frame;
2. Set up a strip of light sensitive film a small distance behind the double slit.

This set-up allows the light to go through the double-slit and imprint its pattern onto the film.

The result on the film, of the light passing through the double-slit, is an alternating dark-light band, which is called an interference pattern.

Figure 1: Double Slit & Interference Pattern. Image by: Jean-Christophe Benoist

To realize the cause of the interference pattern, imagine if instead of light, that we flowed water through the double-slit. The waves going through both slits at the same time bounce off of each other as they exit the slits. Some waves become larger as a result and some become smaller. So the interference pattern is a pattern of wave peaks and troughs. The exact location of each wave peak and trough can be mathematically calculated, and so the interference pattern can be mathematically predicted. The result of waves going through a double-slit of this nature will always lead to an interference pattern similar to the one in the figure above and the one below. Simple stuff so far. ☺

Figure 2 below: With waves, interference pattern forms at top of image. Double-slit is at bottom of image.

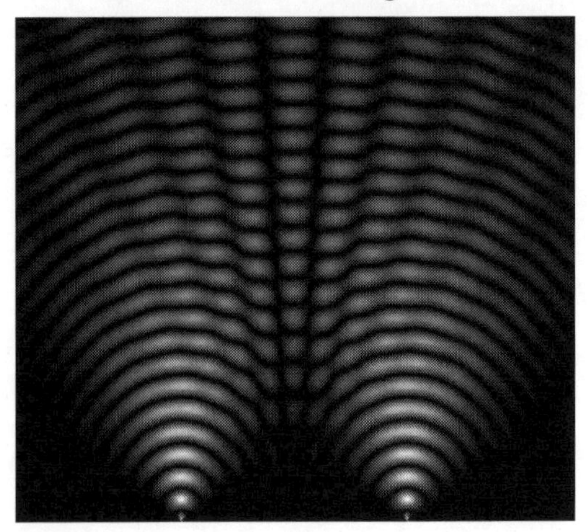

Figure 3: Interference Pattern forms at top of image. Double-slit is at bottom. Image Courtesy: Timm Weitkamp. (CC BY 3.0 DE.)

Now let us go one step further. The next step in this process, moves on beyond Thomas Young's experiment, to 20th Century work. Scientists test the result when physical matter: electrons, atoms, and some compounds, are fired through the double-slit.

The question we must ask in advance is 'what sort of pattern will form when we fire solid particles (like electrons[D3]) through a double-slit?'

D3 - An *electron* is an elementary particle that is a key part of all physical matter.

Imagine the electrons as tiny paintballs being shot through the double-slit, and then hitting the screen on the back wall before falling into a pit below. The next image shows the set-up and the typically expected result of the experiment.

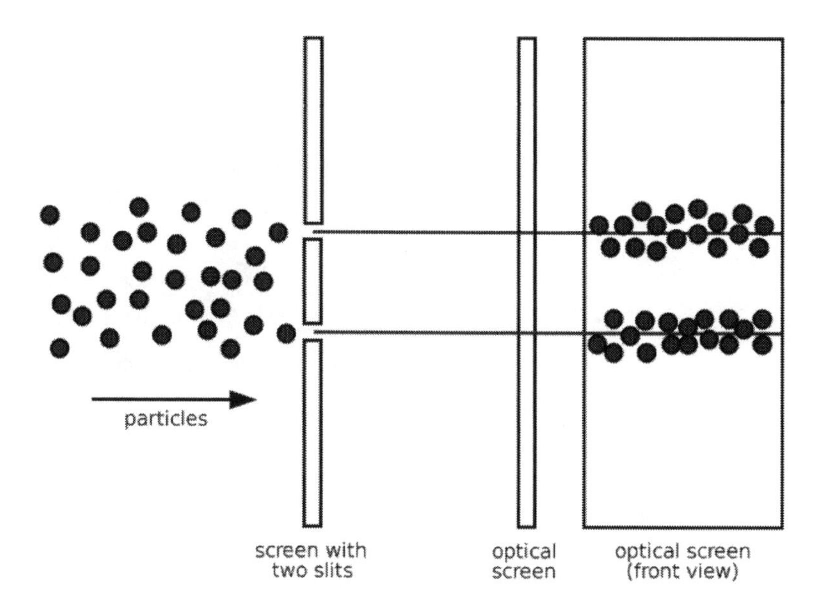

Figure 4: Setup of double slit experiment with subatomic particles and typically expected result.

After firing these subatomic particles/electrons ('paintballs') for a while, we would expect to see the electrons forming a solid double-line pattern on the screen. The pattern should have a resemblance to the double-slit that the 'paintballs' were fired through. There should be two vertical lines of imprints on the film, just like in figure 4.

But, no. We end up with electron imprints scattered in a uniform pattern all over the back wall. Shown next in Figure 5.

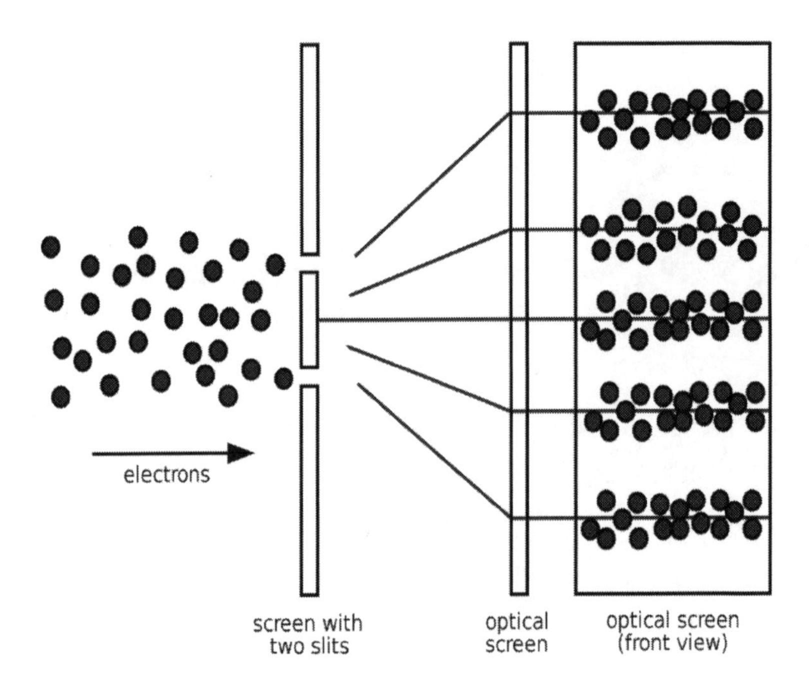

Figure 5: Result of firing subatomic particles, like electrons through double-slit.

Scientists were at first, stunned. This pattern is not what was expected but most interesting is that this pattern is one you have seen before. We realize that shooting single electrons through the double slit has led to an interference pattern. So this means that these electrons we fire are behaving like waves. And the experiment works not just with electrons but with all subatomic particles.

Now this is odd, because our physical body and the entire physical universe is made up of subatomic particles. So for solid particles to behave like waves was a shock, but hang on a bit because the quantum weirdness is just getting warmed up.

Scientists thought that maybe firing all the particles together was causing them to bounce off of each other and behave like waves. You can imagine different paintballs colliding as they exited the double slit and maybe somehow creating a wave like pattern. This is a bit of a stretch but to verify the results the experiment was changed to account for that possibility.

The change, was that the particles were fired one at a time. One at a time, with the electron gun waiting until a particle hits the screen before firing another particle. So the new method using the same setup is:

1. Fire particle at double slit.
2. Wait for particle to hit screen.
3. Fire next particle at double slit.
4. Repeat 10,000+ times.

. . . and the result was: an interference pattern.

Odd, right? Particles fired one at a time created an interference pattern, that is, a wave pattern. The eventual interpretation was inescapable but counterintuitive: by some means, the single particle was going through both slits and interfering with itself to create a wave pattern. The particle was in two or more places at the same time – this is the only way it could interfere with itself.

Nobel Prize Winner for Physics in 1922, Danish Physicist Dr. Niels Bohr:

"Anyone who is not shocked by quantum theory has not understood a single word."

The double-slit experiment is one of the most repeated, retested and verified experiments in quantum physics. Upon encountering it, every student tries to attach a normal explanation to the quantum madness and so too every scientist, but the experiment, even when varied to rule out skepticism has held up for nearly a century.

You have covered good ground so far, but we are not yet done with this quantum insanity.

Activity 2:

1. Grab some paper towels or any absorbent towel.
2. Read the rest of the chapter.
3. Clean your cerebral cortex off of the ceiling.

Now let us continue: You just conducted that quantum experiment, and we concluded that each particle, fired separately, must be somehow going through both slits and interfering with itself - the single particle is going through both of the double-slits. So scientists, not satisfied with this weird behavior of the single particles, set up a measuring device at each of the slits. Imagine a camera type device that monitors the double-slit. If the particle goes through either slit one, or slit two, we can now know exactly which slit was used. Then the experiment was run again with the subatomic particles.

What did we see on the back film when we observed each slit to watch for the particle?

An interference pattern is what we have now come to expect – but, no. The result when we observed the double slits was two solid impact strips in line with the double slits.

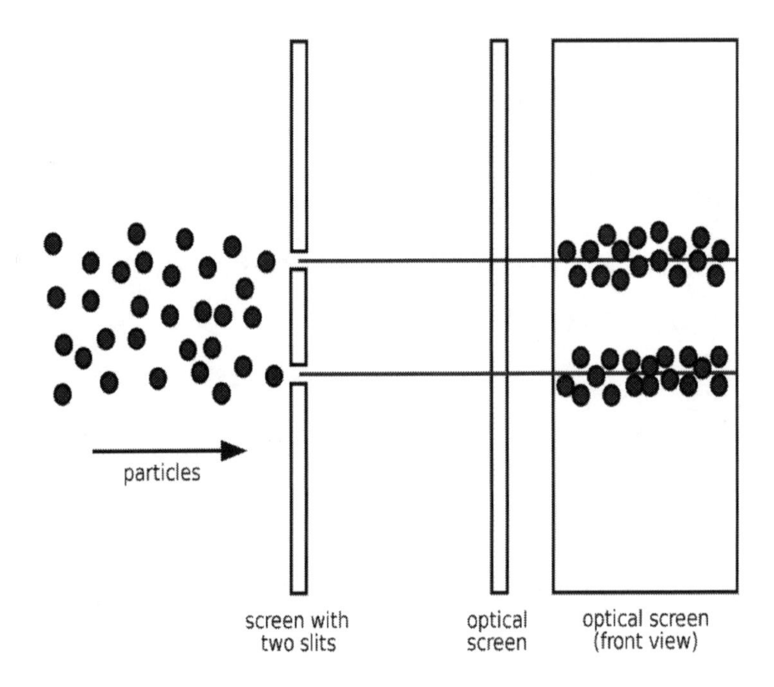

Figure 6: The result when the slits were monitored. Particles behave 'normally.'

Somehow, the electrons altered their behavior because of being observed, and acted, well . . . 'normally.' :-/

Now you understand why this experiment has been so often repeated. You might even Google® 'Double-Slit Experiment' to see video on this experiment. Every quantum physicist has tried to poke holes in these experiments, and run these experiments hundreds of times each, and remodeled it and the conclusion that has become weird but inescapable is that:

The particles that are the very building blocks of all things, are in all possible places until observation causes them to choose a specific position. When a particle is in many places at once the only information that can be accessed is the probability of where the particle might be in a wave of possibilities. This whole conundrum is called 'the measurement problem.'

To quote another Nobel Prize winning physicist, Werner Heisenberg:

> *"The atoms or elementary particles themselves are not real; they form a world of potentialities or possibilities rather than one of things or facts."*

Experiments have gone further with larger and larger particles engaging in this 'superposition,' that is, being in many locations at once.[43] This idea of particles existing simultaneously in multiple places is not just a theoretical concept but is becoming increasingly relevant in practical use as the very early stages of quantum computing are now in progress.[44]

So we must consider that the very particles our bodies and our physical world are made up of can be in multiple places simultaneously and can be configured in an infinite manner of ways, until the act of observing them brings them to one specific place. The very building blocks of our world are in all places and exist as a wave of possibilities until we observe them. This defies our normal understanding of the world because when we look around we see particles, matter, objects in one fixed place and not as a wave of possibilities. To look around is to observe, to observe is to measure, and to measure is to shift all particles from a wave state into fixed positions. To look around is to 'collapse the wave function.'

We can consider taking this idea even further and consider that the double-slit experiment suggests that matter is somehow aware that it is being observed. Notice what happened when the observation of slit 1 and 2 was added to the experiment. We can push the idea to the limits and propose that all reality is just undetermined potentialities behind us, and then the act of our swiveling around and engaging in observation collapses every particle to a fixed position. As though all reality was playing a silly childhood game of 'statues' or as though we were avatars in some advanced version of The Sims®, or as if we lived in what can be called a 'Holographic Universe.' Curious.

> *"There is compelling evidence that the only time quanta [electrons] ever manifest as particles is when we are looking at them. When an electron isn't being looked at, it is always a wave."*
>
> *-- Michael Talbot, The Holographic Universe*

WEIRD REALITY

The more we delve into quantum mechanics the stranger the world becomes; appreciating this strangeness of the world, whilst still operating in that which you now consider reality, will be the foundation for shifting the current trajectory of your life from ordinary to extraordinary. It is the Tao of mixing this cosmic weirdness with the practical and physical, which will allow you to move, moment by moment, through parallel worlds to achieve your dreams.

This nature of the world as different than we have come to know, is captured by the words of Albert Einstein:

"*Reality is merely an illusion, albeit a very persistent one.*"

Quantum physics shows us that in discussing our reality we can only speak of possibility and probability. Just as the future is to us uncertain, your very next moment is in fact uncertain. There is the potential for a particle of matter to be located where we expect it to be or to be located anywhere in the physical world. So it is with our lives - our very next moment can occur along the most probable path or it can occur on a path entirely discontinuous with the expectations that others have for us, but more likely in line with the expectations that we have for ourselves.

We have seen that there is this wave of possible positions of particles and so too we also embrace that there is a wave of possibilities and probabilities for the next moment of our life. As you recall, quantum physicists have called the mathematics of this wave, the 'wave function,' and they call the moment of awareness, when we 'observe' our environment, the 'collapse of the wave function.' Each moment in our lives we shift along with the 'wave function' of possibility, and then as we become aware and break the flow of experience, we collapse the wave function, and our reality becomes clear. With all outcomes being possible, we must start to wonder exactly what determines how the wave function of our life is collapsed into one outcome or another? And the answer to that question is one of the many powerful ideas that you will receive from this book.

We know that observation, that is, awareness, collapses the wave function, but in this work we interpret wave function collapse as described by the 'Many-Worlds Interpretation' of Hugh Everett III PHD. We come to realize that at your point of awareness, your world splits. The collapse of the wave function establishes us into a specific world and then we continue on in the flow of that world. We shall explain this now in the 'Many Worlds Interpretation of Quantum Mechanics.'

MANY WORLDS INTERPRETATION OF QUANTUM MECHANICS

We now know that a single particle can exist in two or more different locations at the same time. The particle exists in multiple locations, and has effects in multiple locations as a wave of possibilities. The exact location of the particle if observed or measured is determined by probability – the particle may be in the location where it is most expected or it can theoretically be at any place in the physical universe. German-British physicist and mathematician Max Born, in 1926 called this a 'probability wave' or 'wave function,' and the terms remain today.

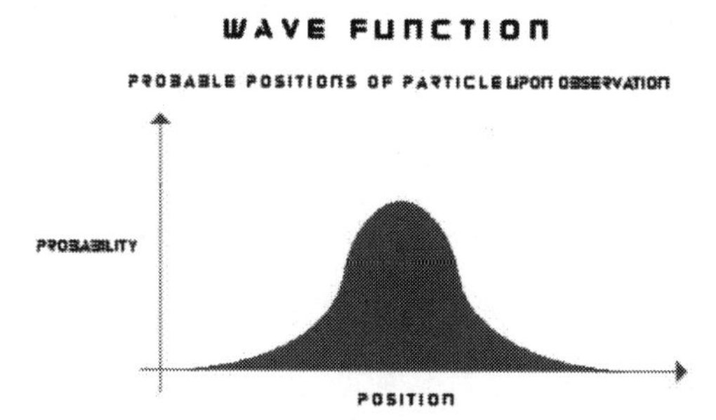

The particle, as we now know, is said to exist in superposition (two or more places at the same time). This wave existence of the quantum particles that make our world raises significant issues. If by observing a quantum particle it suddenly goes from being in multiple places at once to revealing itself in just one place, then what happens to all the other probabilities or existences of the particle that are not realized? We look at a particle and it appears only in one place but without observing it, the same particle existed in multiple places as this wave. So what about the other points in the wave where the particle could have been?

This question is famously raised by legendary Austrian physicist Erwin Schrödinger, in a thought experiment where he describes the fate of a cat in a box as being dependent on whether a quantum particle is in one location that triggers 'the bomb' that kills the cat, or if that particle is in another position that disables the bomb. Schrödinger suggesting that with the quantum particle being in both the trigger and non-trigger positions until we observe the quantum particle, then the cat must also be in a state where it is both alive and dead until observed. The takeaway here being that quantum mechanics seemed to be suggesting that two or more contradictory events can be occurring simultaneously.

In 1956, Princeton PH.D student, Hugh Everett III, wrote a paper originally titled: the 'Theory of the Universal Wave Function' which describes the detailed mathematics of what occurs when we observe the single location of a quantum particle that was previously in superposition. That is, the paper described what occurs when our observation causes the wave function to collapse to a single observed reality. The mathematics of the theory led to a conclusion that upset the current establishment and that was even more counter intuitive than what we have seen so far in quantum mechanics.

Figure 7: Schrödinger's Cat; Many-Worlds Interpretation, with universe splitting.

The mathematics of the theory suggested that when one particle behaves as if in multiple places at once, it is in fact only in one place in our world but in the other places in other worlds identical to our own (identical, except for the impact of that differing particle location). So in one world Schrödinger's cat is alive and in another world Schrödinger's cat is dead. In one world you muster the courage to approach the girl/guy of your dreams, and live happily ever after . . . and in the other world you do not.

By observing or acting on the particle in any one specific location, we become a part of the world that has the particle in that location. Stated differently; if one particle is at A and B at the same time, then there are two worlds that are overlapping – the world of A and the world of B. To observe or measure or be aware of the particle at A is to enter into the world of A exclusively, and to cause the world of B to cease to be accessible to the observer. The world is splitting into copies of itself every time the process of measurement or observation, that is, consciousness, interacts with quantum elements. Stated differently, every time we observe our physical world, which is comprised of quantum elements, we are splitting reality into different paths and trajectories. Both metaphorically and literally, our consciousness and awareness, which includes our choices on what to place our focus on, alters the paths of our lives and creates a split in reality leading to creation of two or more worlds at any moment in time. Worlds that are very real, but separate from our current ability to consciously interact with or influence.

In the Many Worlds Interpretation of Quantum Mechanics the trajectory of your life is no longer just one straight path to an eventuality, but is instead one path of many on an ever-branching tree of possibilities.

POPULARITY

Informal polls have shown that more than forty percent of quantum physicists believe that this Many-Worlds Interpretation is literally correct. Legendary quantum physicist Stephen Hawkins is said to be among those who stand by this theory, as well as Nobel Laureates Murray Gell-Mann and Richard Feynman - admittedly all three have reservations about its title 'Many-Worlds' and the physical implications of the theory, but not about the basic structure of the theory. Polls do not determine science, but I include this point to highlight the popularity of the theory in spite of its metaphysical and practical consequences.

In the book, The Physics of Immortality, by Frank J Tipler, the author highlights a pre-1988 poll conducted by L. David Raub, of

72 leading cosmologists and quantum field theorists that places the numbers thusly:

1) "Yes, I think MWI is true"	58%
2) "No, I don't accept MWI"	18%
3) "Maybe it's true but I'm not yet convinced"	13%
4) "I have no opinion one way or the other"	11%

The Many Worlds Interpretation can be seen to suggest a constant creation of new physical worlds – which is what the mathematics suggests, or it can be viewed as worlds only occurring in the mind of the individual. The occurrence of these worlds in the mind does not suggest that it is not physical – although this may sound like a contradiction. The existence of a reality in the mind does not cause it to be any less real. I state this because we must realize that the entire world exists only in our minds. All that we perceive and feel and see is the result of electrical impulses in our brain. The world of the individual is tantamount to, or potentially quite literally, a highly advanced computer running and analyzing programs in its working memory. That is, your mind is just a computer processing information and everything that seems physical is electrical bits interpreted by the brain. I do not suggest that you must make the leap now and embrace this view of the self as non-physical, nor would embracing this view diminish the importance of the soul or make human existence any less real - but it does suggest that we all would do well to consider that our intuitions about how the world works may be incomplete. Quantum physics has shown that when we break down the constituent parts of matter we do not get solid substance but more than 99% of every physical object is made of empty space. Empty space with a buzzing array of quantum particles bouncing in and out of reality. Just an abundant soup of protons and electrons – think electricity, think bits, think 1s and 0s. There could not be a more apparent display of how unphysical the world is than that bit of scientific revelation. We live in a world of bits and bytes - from this perspective, which I appreciate you may be reluctant to embrace at first reading, we start to consider that parallel worlds that allow us to jump from one path to another in a continuous or discontinuous manner are not that hard to fathom. And even if it is hard for someone to fathom, I wonder if reality cares? Reality does not exist to make any one of us feel comfortable - in fact, no matter what one believes, we can agree that we exist in this reality for the purpose of growth - be that spiritual or evolutionary. And growth comes not from comfort, but from discomfort. So I suggest that you are able to embrace the discomfort of an altered view of reality as through this discomfort and growth of perspective comes the opportunity for further growth. We must allow our minds to behave like the very quantum particles we discussed - our minds should be able to have ideas that are different or contradictory but still existing simultaneously. We must be comfortable having our view of the world itself existing in superposition.

> *"The test of a first rate intelligence is the ability to hold two opposed ideas in the mind at the same time, and still retain the ability to function." - F. Scott Fitzgerald*

Throughout this book I will offer you ideas that can be accepted as either literal, metaphoric or both. The ideas are both things. The ideas are each thing. The ideas are all things. Some ideas are no thing. The ideas are all in superposition and thus your takeaway from this book is unique. Remember, you are co-creating this book.

THE MATHEMATICS OF THE THEORY

The Hugh Everett III 'Interpretation of Quantum Mechanics' now called Many-Worlds, is mathematically solid, and is consistent with the science of quantum mechanics - indeed it should be the default view of all scientists based on this. The reason it is not the default view of quantum physicists is because the implications of the theory are disconcerting and present a very odd view of physical reality. Some prominent scientists have even stated a dislike of the theory because of the metaphysical (spiritual) implications (for the metaphysical implications, see this book ☺).

My job in helping you achieve your dreams is to allow you to use the paradigm and implications of the theory to your advantage. Physical reality does not require that we be pleased with its mechanism and we must see the implications of a theory for what they are and not for what we would like them to be. Metaphysically these ideas are challenging, but what believer of faith among us can claim to understand the exact mechanistic structure of a world created by a god/God. Conversely, I suggest the secularist should be very comfortable with the math of The Many-Worlds Interpretation and should embrace it because of its mathematical rigor in spite of its metaphysical implications.

EINSTEIN'S LESSON

The mathematics of Albert Einstein's theory of relativity suggested that the universe was expanding and doing so at an in-creasing rate, but this idea was so outrageous at the time, that Einstein rejected it as a practical idea, and created a concept called the cosmological constant to counter what the math was showing. When finally Edwin Hubble showed that indeed the Universe was (and still is) expanding and doing so at an accelerating rate, Einstein called his own personal refusal to accept what the mathematics was showing as: "the greatest blunder of [his own] career." To ignore the sound mathematics of a theory like Many-Worlds because the result is uncomfortable, is indeed, to blunder greatly.

AN INFINITY OF POSSIBLE WORLDS

There are many, many, many worlds branching out at each moment you become aware of your environment and then make a choice. One may think that an almost infinity of worlds branching off from each human observer would be all too many worlds, and all too much matter to create. I respectfully suggest to you that this is based on two human delusions you still have:

1) That the world around you is a physical construct that is not occurring in your own mind, and

2) That all other persons on this planet are separate, and that one's own consciousness is not directly connected to and entangled with all other consciousness and all 'particles' in physical reality. That is, you the reader believe that they you are not omnipresent, omniscient and ultimately omnipotent - you may believe that you are separate from that which you would describe as god.

"Jesus answered them, Is it not written in your law, I said, You are gods?" - John 10:34

"I have said, You are gods; and all of you are children of the most High." - Psalm 82:6

To state those two points in a more secular manner:

1) We are not all aware that we live in a 'Holographic Universe,' and

2) We fail to realize that all our particles are engaged in a form of quantum entanglement with all other particles.

In either case, you need not embrace these two ideas to benefit greatly from this book - I am merely placing those two perspectives out there for your later consideration or realization. In the next few decades research will continue to build regarding the concepts of the holographic universe and quantum entanglement, so consider these two points as an inside scoop.

YOU ARE ALSO REPLICATED

In Many-Worlds, even your own consciousness is replicated at the point of awareness. The idea that there are infinitely many versions of the self, existing in many worlds and seemingly separate from interaction, is in line with many common spiritual teachings. The word 'Maya' suggesting 'life as mere illusion' is referred to in the Hindu, Buddhist and Sikh faiths. If we look upon the earth as a place where our 'higher selves' have come to learn, to experience or even to be judged, then the splitting of realities is merely an extension of these functions.

OPPORTUNITY

The Many-Worlds Interpretation speaks to possibility and it speaks to opportunity. By appreciating its existence and adopting the paradigm of its existence, we start to realize our future has infinite potentiality, and we realize that our **Ideal**

Parallel World (IPW) exists already in one path of our potential future, and therefore our behaviors in the present can guide us there.

"So many worlds, so much to do . . ."
- *Alfred Tennyson*

So we have started this book by allowing the reader to consider that they are in fact directly connected to god and to all things; and from considering this paradigm and applying it with humility, life proceeds with a much greater ease.

THE HOLOGRAPHIC UNIVERSE

The Holographic Universe is a currently growing theory in physics that suggests that the 3 dimensional universe that we live in (four dimensional with time), is actually a projection of the two dimensional surface of a black hole. To describe it simply, one can imagine the black hole as a projector that surrounds us and gives us the illusion of the three dimensional world (plus the fourth dimension of time) that we perceive. As with much in quantum physics, the idea is counter-intuitive but when the math and the research suggests that reality is operating differently than we perceive, it is wise to suspend our disbelief and consider that the mathematics and scientific observations should be taken seriously in spite of how we may feel.

One of the keys to this understanding of the universe was the observation that items taken into a black hole have their energy dispersed into space as radiation but that the information contained about the item is stored on the surface of the black hole. There was observed to be a direct relationship in the amount of information contained in the item entering the black hole and the size of the exterior surface area of the black hole. That is, as the black hole sucks in, a star, for example, the surface area of the black hole increases exactly enough to represent the full dimensions and every single detail of the star on the black hole's surface area. So, if to represent the star we need a quintillion bits of data, then the surface of the black hole increases by exactly enough space to store a quintillion bits of data. This is in fact a fundamental law in physics that information as to the state of a system is never lost, and therefore information about any object within the system is also never lost. In such, the information needed to create any/all objects that has ever existed is immortal.

I have introduced this concept of the holographic universe in this work, not so that you may fully embrace the idea. But merely to continue to awaken the reader to the possibility that the world may operate in ways that are counter-intuitive.

Again, none of this changes the reality of our existence, or our relevance as conscious beings, nor does it rule out the existence of god, but it does suggest that our perceptions as to reality, and as to how the universe operates, structurally, may be flawed.

For more on this, read the 'Holographic Universe' by Michael Talbot.

MANY-WORLDS AND MANY-MINDS

Jeffrey A. Barrett Ph.D., Columbia, 1992, has written a brilliant paper about the 'Many-Minds' approach to quantum mechanics. In that paper Barrett suggests that these parallel worlds exist primarily in the mind of the individual; with the conscious mind having access only to the one world we end up in.

CONCLUSION

So to view the world as non-physical, to view matter as not existing separate from your own awareness, and to consider the existence of many worlds branching out at each moment because of awareness, are all uncomfortable ideas. One need not even embrace them literally, as they are just as effective considered metaphorically. But all this, is a launching point.

In regard to Hugh Everett III's theory for a moment, the mathematics is very clear, and the theory stems from the mathematics, not the mathematics from the theory. It is truly a remarkable piece of work. It causes us to consider that there are worlds emerging in superposition with our own that we are not yet able to directly access. This book is about bringing you access to this infinity of superpositioned worlds, by altering your paradigms, and applying 'The 2 Pillars of Parallel World Travel.'

STAGE 3

Creating the Vision (Your NIF and IPW)

"The greatest danger for most of us is not that our aim is too high and we miss it, but that it is too low and we reach it."

— Michelangelo

THE FRAMES OF OUR LIVES

Every second, atomic clocks, powered by the cesium atom, 'tick' more than 9,000,000,000 times. Located throughout the world in military and high technology installations, atomic clocks set the international standard for time. The two most important clocks in the United States are located at the National Institute of Standards and Technology (NIST) in Gaithersburg, Maryland, and the second clock is at the U. S. Naval Observatory (USNO) in Washington, D.C. Readings from the clocks of these two agencies contribute to world time, called Coordinated Universal Time (UTC). The most accurate atomic clocks invented, lose less than a second every ten billion years. So that will do.

Consider that every hour of your own life that ticks by, there are numerous points and moments when you can choose to pause - and be aware of your mental and physical state. At any given point you can freeze the flow that is your life and be fully conscious for a moment. At any given point, you can pause, take a breath and take a snapshot of your reality. At any given point, you can look around you and acknowledge the frame, in the film that is your life, and become aware of what you are thinking and feeling in that frame.

When we enjoy a film, all we are really watching is a series of still frames, one after another. At the movie theater we usually are watching films projected at twenty-four individual frames per second. This is more than enough to give the brain the perception of movement, even though we know it is indeed a series of still frames that are being shown. One of my favorite movies, 'The Hobbit,' directed by Peter Jackson was initially shot at forty-eight frames per second and some initial screenings suggested that this high frame rate 'broke the illusion of fantasy' because it was 'too lifelike.' The suggestion being that as little as just forty-eight frames per second was crossing the boundary from fantasy to reality. The brain has evolved to sense and anticipate movement, and with this, an attribute called 'persistence of vision,' in the brain, gives us the sensation that snapshots of reality represent a kind of movement and correspondingly a passage of time. We see our reality unfolding one frame after another because we have this perception of the disparate frames as part of the flow that is our life. It is the advance of each of these frames that moves us ultimately to our Ideal Parallel World - which we will shortly discuss.

If we considered that the holographic world we live in were merely frames, then the frame rate would be closer to a 9,000,000,000 frames per second, on par with the behavior of the cesium atom. Perceiving 9,000,000,000 frames per second does not seem achievable for the human mind - maybe for a more evolved entity it could be. However, it is the point when we become aware of what we are thinking, feeling and doing in any given moment, that is, it is at the point of awareness and consciousness of the frame we find ourselves in at a given instant - it is at that precise point, that we split reality. It is at a point of choosing one thought over another that we split reality. It is at the point of embracing one idea over another that we split reality. The quantum particle in superposition is an electron existing in our very mind at the point of becoming aware, for a single moment and single frame. Choice, is what presents us with a multitude of paths because choice creates a flow of electrons through the brain in a manner that inexorably leads to quantum superposition, and the many worlds that are the inevitable result.

The paradigm I propose to you, is the perception of your life as this series of frames. So many great teachers have taught humanity that it is in consciousness and awareness that we evolve. That it is at the points in our life when we choose a path of awareness, that we make our greatest strides of personal growth. Awareness splits reality and allows us this choice between parallel worlds.

> *"The key to growth is the introduction of higher dimensions of consciousness into our awareness."*
>
> *- Lao Tzu*

When we make the choice to be aware of what we are thinking and feeling in each moment – we bring in the region of the brain responsible for reason, rationale and whole brain integration (called the Pre-Frontal Cortex). This awareness gives us more control over the moment, over ourselves. This awareness freezes the flow of existence and allows us the perception of the frame we are in, and awareness of the frame is the opportunity to shift the flow and to start the movement between those parallel worlds.

When you use the power of consciousness, moment by moment, you are shaping and bending light to create a future of greater possibility. You are shaping light – you are bending light – altering forms and expressions of light to create new possibilities, new realities.

This book works to help you shift the trajectory of your life from whatever straight line of existence you find yourself on, to a path chosen along a branching tree of possibility.

Through consciousness, one has the power to surmount any failings of the past. Failure, is being stuck on a predetermined path unpleasing to the self. Success is breaking free of that unfulfilling destiny and progressively switching through parallel worlds within your field of probability. Like switching between channels, every moment we allow ourselves to be aware, we have the opportunity to switch worlds and to split tape, and move the film of reality over to a scenario more to our liking and closer to our Ideal Parallel World(IPW).

Legendary life-coach and author of numerous international bestsellers, including my personal favorite 'Awaken The Giant Within,' Anthony Robbins, tells a fantastic story about his golf game. He mentions that if the point of impact of the golf club on the ball is just one millimeter off center, then that extremely small difference can mean the difference between the ball landing in the ocean or landing perfectly on the putting green.

In string theory, this is called 'the law of exponential sensitivity to initial conditions of chaotic systems,' and suggests that small shifts in initial starting points can lead to dramatic changes in the end result. It is the result where you shift the trajectory of your life in any one area by just one degree, how that small alteration is magnified as time progresses. The difference between making one million dollars a year and living like a prince versus $30,000 and just being able to survive, is often as little as a one degree shift when magnified over five to ten years. A one percent increase in your brain power from using the 2 Pillars, can lead to a significantly more fulfilling and successful life - but I suspect, as someone reading this book, that you are capable of achieving much more than just one percent growth.

As you consider the law of exponential sensitivity to initial conditions, realize that ultimately your end point will be vastly different than the point where you started - you will be in a literal and/metaphoric parallel world. So as you start to imagine what the world of your dreams looks like, be open to all possibilities and even improbabilities. Small shifts in your thinking and small shifts in your energy, can lead to massive changes in end results, so embrace your responsibility to dream big.

"Ask and it will be given to you; seek and you will find; knock and the door will be opened to you."

- Matthew 7:7

Now, you must become more clear about what you intend to achieve, and begin to have a knowing of exactly where you would like to be in your life. You shall now begin to think of what I call your Next Ideal Frame (NIF) and you shall also begin to think of the precise vision of your Ideal Parallel World (IPW). The image you create of what your Ideal Parallel World will look like will be fundamental to everything else that you do. You must see clearly the person you will be in that IPW, you must decide with specificity what assets you intend to possess and the type of occupation you will have or even better the type of business you will own, and as specific as possible, the character of the person you will travel to become. This is about clarity, yes – clarity of vision and also clarity of emotions.

Your Next Ideal Frame (NIF) is a goal that is close enough to where you are now that you can chart a course to your NIF using your conscious mind. The Next Ideal Frame is a 'frame' as we defined before, that exists approximately 3 months from where you are right now (you may choose a sooner frame but only if you really have a lot of free time). This frame 3 months from now is a parallel world, yes, but it is one that is closer to your current frame than is your ideal parallel world. So if you are a salesperson – increasing your sales by fifteen percent when you were only initially shooting for five percent would be a solid NIF. It is a real parallel world because it is more than you would have achieved on your current trajectory, but it is not meant to be a shockingly different world from the one you are in now. The reason for this thee month NIF is to give you time to master the '2 Pillars' in this book before you really step on the gas with your parallel world travel. If you have a lot of available 'frames of awareness,' that is, a lot of free time, in the next forty-five days then you can set your NIF to a forty-five day point.

So to describe the NIF using our salesperson as the example:

A Next Ideal Frame should be close enough to your current reality that if you sat down right now, you could create a strategic plan for getting there that would describe the emotions needed to get there (joy, happiness, passion, etc.) as well as the behaviors you will need (meeting with better sales mentors, pursuing bigger leads and ignoring smaller leads, hitting the gym every day to improve your appearance, etc.). The NIF is a point closer to now than is your Ideal Parallel World. It is a <u>tiny</u> sampling of your dreams.

Now grander, your Ideal Parallel World (IPW) is everything. Your IPW is your dream of what you intend to achieve in the future and should be so grand that you could <u>not</u> chart a course to your IPW if you tried to right now. Your IPW must be so grand that using your conscious mind, you could <u>not</u> chart the course to it – if you can chart the course to your IPW with your conscious mind right now, then it is not big enough.

Your IPW must be so grand that if you told it to 99% of your friends, they would laugh – some because they would think you were joking and some because they would think the goal impossible. That is fundamentally why this book is about moving between parallel worlds to achieve your dreams, because the point of achievement must be so different from where you are now, that achieving it would seem to others like you have stepped into another reality. Indeed, the IPW is a goal so grand that the only path to achieving your IPW is to travel through parallel worlds and this will require the 2 Pillars we will discuss. Over time, by using the 2 Pillars, your subconscious mind will give your conscious mind the strategic plan for getting to your IPW, and the subconscious mind will also cause you to find pleasure from doing all the behaviors listed on the strategic plan you will create (your subconscious mind will cause you to love walking up to successful people and convincing them to mentor you, or cause you to love waking up at 5am to go to the gym).

"Imagination is more important than knowledge."

- Albert Einstein

If you are one of the two billion or more people who believe in the teachings of Christ, you must know that we all have the power of God in us. That we are all made in His image and like Him we can achieve miraculous feats. Jesus said to his disciples:

"Truly I say to you, If you have faith, and doubt not, you shall not only do this which is done to the fig tree, but also if you shall say to this mountain, 'be you removed, and be you cast into the sea,' it shall be done."
– Matthew 21:21

I would ask all believers, why would Jesus say such a thing if it were not true. If you believe in Him then you know it must be true. We must have the power to speak and have matter yield to our demands. And most certainly we must think highly of our potential power and see it as our responsibility to live up to the fullest of that potential. That is what the Ideal Parallel World (IPW) is all about. It is the place where all that we have imagined is in place. It is a point from which we are still creating and manifesting and accomplishing and still have much work that is to be done, but it is a place that we would like to be at, in the present moment. If you aspire to be a great CEO, then in the vision of your Ideal Parallel World you can see yourself having just been voted in as CEO by the company's board or just taking a company you founded to the point of its Initial Public Offering of stock. From there much work lies ahead but this is a point where you would like to see yourself and a point where essentially you wish to be at right now. Once you get to the IPW, and you will, you then will either create a new, grander IPW or will create specific Next Ideal Frames to keep you growing.

CREATING YOUR IPW

> *"If one advances confidently in the direction of his dreams, and endeavors to live the life which he has imagined, he will meet with success unexpected in common hours." - Henry David Thoreau*

Spend some time reflecting upon and then return to answer the following questions after you have completed this chapter. These questions are not meant to cover all aspects of your IPW. The questions are meant to get you started in the process of creating your IPW. Once you fully understand how grand your IPW is you will be ready to begin creating this image of it. You may scribble in some brief answers on here in pencil, or you can do so on a separate sheet of paper, if you do not want someone else to pick your book up and laugh at your IPW :-)

Here are some questions:

What is your net worth at your IPW? (Decide on an exact net worth - you can give some money away if you end up with more than you planned. ☺)

List the most important/necessary/useful assets that you have in mind (house, car, property, businesses)

Decide what great service, product or skill you will offer to humanity that has created this wealth? Be specific.

What relationships are strong and important to you in the IPW (including, family, friends and romantic)?

What behaviors and virtues do you display at your IPW that are key to your continued success? (You will have to edit this answer after reading about the '2 Pillars.')

What are your dominant emotions at your IPW? How do you feel about life? (You will have to edit this answer after reading about the '2 Pillars.')

Anything else? Health, Wealth, Social, Business? Be specific.

Activity 3:

Next, in creating the image of your IPW, you should gather up some actual images from the internet, magazines or other sources available to you. Images that represent the type of luxury existence you will have at your IPW and save these to a device or folder you can access regularly. (You may also create what is commonly called a Vision Board. Search online for examples of 'Vision Boards')

Memorize the look of these images in your conscious mind, so that when you have a moment of peace, meditation or self-hypnosis you can plant them into your subconscious mind. (We shall discuss how to do this in Pillar I.)

That is it. You are on the right path.

"I bargained with Life for a penny,
And Life would pay no more,
However I begged at evening
When I counted my scanty store;
For Life is a just employer,
and gives you what you ask
But once you have set the wages,
Why, you must bear the task.
I worked for a menial's hire,
Only to learn, dismayed
That any wage I had asked of Life,
Life would have willingly paid"
- The Wage, Jessie B. Rittenhouse

CREATING YOUR NIF

To create an image of your Next Ideal Frame, you are using a highly diluted version of your Ideal Parallel World and targeting a point about three months from now. This three month time period is necessary for you to master the 2 Pillars. The three month period is not a limitation, as in theory you could move over to your IPW instantaneously - although this is by definition, highly improbable. (Also, if you do create an IPW which can be manifested in less than three months, then I would suggest you need to dream a bit bigger.)

Do not create your NIF until you are complete with the entire book. This is because you will need to incorporate what you learn in the 2 Pillars into the Next Ideal Frame (NIF).

When you do complete your Next Ideal Frame (NIF), you will be answering some basic questions that you may fill in here or in a separate journal:

In three months, what will your life look like in terms of:

a) Your Health

----

b) Your Wealth

----

c) Your Relationships

----

d) Your Career

e) Your Personal Growth

f) Your Emotions

g) Your Behaviors & Habits

h) Other Key Goals with Dates of Achievement

To get to your Ideal Parallel World (IPW), you move through many Next-Ideal Frames (NIFs) and many-worlds. Each world you move through is a stage along a greater process. In every moment when you choose, you create a new destiny, so there are multiple destinies unfolding -- multiple parallel worlds that you must pursue along your path -- each with its own images, thoughts, emotions, behaviors and habits that you must embrace.

When you complete the 2 Pillars and have read 'Timeline For Success,' you shall be ready to complete the NIF information above and would basically have the start of an IPW Guide that you should add to a journal:

Page 1: The full vision of your IPW.

Page 2: All the ideas you will need to plant into your subconscious mind in order to achieve success. A deadline for arriving at your IPW.

Page 3: Your full current NIF.

Page 4: All the ideas you'll need to plant in your subconscious mind to get to the NIF. A target date for arriving at your NIF.

This is goal setting.

1. **What** you intend to accomplish (NIF & IPW).
2. **When** you intend to accomplish it.
3. **How** you intend to accomplish it (Strategy & Behaviors: The inspiration for the '**How**' regarding your IPW will be revealed to you, through Pillar I, as you approach your NIF).

Your NIF focus will be most important to ensure that you are making progress towards your IPW. Your IPW focus will be to ensure you stay connected to the infinite power and energy of who you are in your IPW, and that you do not get lost in the flow of reality.

"Where there is no vision, the people perish . . ."

- Proverbs 29:18

THE 2 PILLARS FOR PARALLEL WORLD TRAVEL

To move through Parallel Worlds To Achieve Your Dreams you must use the 2 Pillars:

1. Subconscious Power
2. IPW Thoughts

Each pillar serves a specific function in your journey.

1. Subconscious Power - Provides the map through parallel worlds.
2. IPW Thoughts - Ensures you remain aligned with your IPW, also provides the 'fuel' to travel between worlds; this is your source of power to grow and it is the energy to persist.

Both pillars will be explored and explained, and the methodology for using each pillar shall be outlined. You will master these two pillars and then achieving all things shall be inevitable.

PILLAR 1

SUBCONSCIOUS POWER

"A great and sacred spirit talks indeed within us, but cleaves to its divine original." – Seneca

The subconscious mind knows the path you must take to arrive at your Ideal Parallel World (IPW). The subconscious mind is your guide. The subconscious mind is your intuition. To strengthen the connection between your conscious mind and your subconscious mind is to gain access to a map and compass as you travel across parallel worlds. To gain access to the subconscious mind is to gain the ability to see and create the future, the ability to shift the present, and the ability to alter your own perception of the past. To master this first pillar, is to master one's self. To master this pillar, is to achieve an unfathomable power of creation.

> *"Whatever the mind of a man can conceive and feel as real, the subconscious can and must objectify." - Neville Goddard*

Every second, your brain is taking in millions of bits of data through your physical senses. Moment by moment, enough data to crash a supercomputer is rushing into your brain. Your brain is processing all this data and creating useful information and a unique view of 'reality' for the conscious mind. The brain is making decisions as to what is important right now and what can be ignored for now but saved for later. Even whilst filtering through massive chunks of this data, the parts of the 'subconscious' brain called the hippocampus and the basal ganglia are taking statistics on all the data, and deciphering patterns from the data.

This continuous data processing is occurring outside your conscious awareness, but all within your own brain. This background or what we have called 'subconscious' activity accounts for ninety-five percent to ninety-nine percent of your mental activity when you are awake. Think about that . . . up to ninety-nine percent of the processes being run by your brain whilst you are awake, are outside of your conscious awareness. More than ninety-five percent of the mental activity in your brain as you consciously read this sentence, is being used by your subconscious mind.

The conscious mind is the very narrow area that is your current focus and includes most of what Austrian Psychoanalyst Sigmund Freud called the ego. The conscious mind is the mind that reflects on these words as you read them but can only process about five to nine chunks of information at a time. Some psychologists go even further, and suggest that because the conscious mind has to process information sequentially, even thinking of more than one item at a time is like trying to rub your belly whilst patting your head - it is very difficult . . . unless you practice. ☺

The subconscious mind, refers to all mental activity that is occurring below the point of awareness and at any moment, can process more than two thousand times more information than the conscious mind.

The conscious and subconscious coexist and are housed in your brain. There are specific parts of the brain's structure that can be linked to the conscious mind and parts that we can link to the subconscious mind. We shall explore these over the next few pages. I will use the term brain and mind interchangeably as this is sufficient for our understanding of this pillar.

The conscious mind is primarily comprised of the prefrontal cortex and parts of the cerebral cortex. The prefrontal cortex is a small region just above the eyes, behind the forehead, at the front of the brain. The prefrontal cortex is responsible for integrating all the processes of the brain and for the level of strategic thinking characteristic of humans. The remaining cerebral cortex is a thin one millimeter sheet of brain cells (neurons) that covers the entire brain like a wrinkled blanket.

The subconscious mind, is essentially all other parts of the brain. The key parts of the subconscious brain we shall note are the hippocampus, the amygdala, the thalamus, the nucleus acumbens and the basal ganglia. These five areas are all part of the limbic brain, also called the reptilian brain.

All data received by the body is first processed by the hippocampus, basal ganglia and amygdala before being relayed to the conscious mind. Data is only passed on to the conscious mind if the subconscious brain determines there is a need to do so. The subconscious mind controls most biological functions outside of your awareness; this includes respiration, temperature control, heart rate, digestion, immune system, the blinking of your eyes, responding to cellular behavior, and all other systems that one does not usually control with the conscious mind.

The subconscious mind has the responsibility of taking in a virtual infinity of data and filtering it down to that which must be processed by the conscious mind.

> *"The human brain produces in 30 seconds as much data as the Hubble Space Telescope has produced in its lifetime."- Neuroscientist Konrad Kording, Northwestern University*

The hippocampus and basal ganglia control habits, and the amygdala controls certain powerful emotions like fear. All parts of the brain we have noted are linked to a part of the brain called the nucleus acumbens. The nucleus acumbens produces a powerful chemical in the brain, a neurotransmitter, called dopamine. Dopamine is the main pleasure chemical in the brain. When the nucleus acumbens releases dopamine you feel great and think very clearly. Any behavior engaged whilst dopamine flows through the brain is likely to be repeated and the connection of neurons (brain cells) that triggered that behavior becomes stronger. For example, the act of smiling, releases dopamine; having someone you find attractive smile at you releases dopamine, and sex releases an extreme amount of dopamine. Many stimulant drugs including caffeine, heroin, and amphetamines used to treat ADHD, cause the brain to release dopamine. Dopamine release can also be impacted by your conscious thoughts when we 'decide' to be happy, but it is typically controlled by the hippocampus and basal ganglia. The hippocampus and basal ganglia cause dopamine to be released whenever you engage in an activity that is habitual or any activity that the subconscious mind has determined is key to your own survival and the survival of your genes (the statistics taken by the basal ganglia and hippocampus allow the subconscious mind to make the determination as to what behaviors serve this purpose).

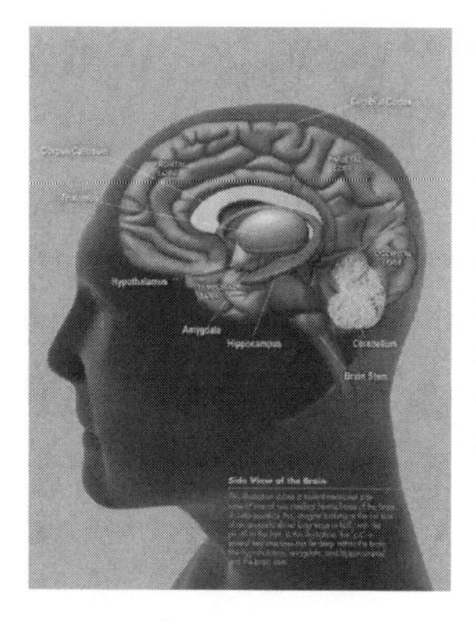

-The thalamus, hippocampus, basal ganglia, nucleus acumbens and amygdala control and create our emotions and drive more than ninety-five percent of our behavior.

-The hippocampus and basal ganglia have major roles in long term memory and pattern recognition.

-The amygdala can activate the 'fight or flight' response.

The thalamus: makes the decision on what information should be relayed for the conscious mind to decide on the next step[45] (for example, the thalamus would not ask the conscious mind whether or not to keep beating your heart or whether or not to stop sweating, but it will ask the conscious mind whether or not to eat a sandwich or cross the street). That feeling we get of deciding to do something is often just us choosing among options presented by the thalamus. This is one of the ways the subconscious mind guides our path through life, our path through many-worlds, and determines what we perceive as reality. Our thalamus decides what stimuli the conscious mind should focus on – it controls our attention. If a stimulus in the environment is novel enough, threatening enough or has the potential to benefit us, the thalamus alerts the conscious mind to focus on it.

THE ILLUSION OF CONSCIOUS CONTROL

As mentioned, the subconscious mind, makes the decision as to what matters get relayed to the conscious mind. The feeling we get of observing reality and making decisions is partly an illusion. When we get the sense of making a decision, indeed we are, but your decision is made from a preselected list of choices compiled by the subconscious mind and sent from the thalamus to the prefrontal cortex. This whole process happens in less than 0.2 seconds. There are numerous experiments that highlight this. Neuroscientist at the Bernstein Center for Computational Neuroscience, based in Berlin, Professor John Dylan Haynes, performed a key experiment that illustrated that the subconscious mind makes many decisions before informing the conscious mind.[33]

The experiment, was done by using brain scans (fMRI) on subjects in the process of decision making. For basic 'A or B choice' decisions, researches looking through an fMRI can actually tell whether the subject will pick A or B as much as ten seconds before the subject becomes aware of the decision they are about to make! Professor Haynes is predicting the person's decision, with more than ninety percent accuracy, before the person is consciously aware of having made a decision. Haynes explains the findings:

> *"We found that the outcome of a decision can be encoded in brain activity of prefrontal and parietal cortex up to 10s before it enters awareness. This delay presumably*

reflects the operation of a network of high-level control areas that begin to prepare an upcoming decision long before it enters awareness."

So Prof. Haynes' conscious mind knows the decision before the subject's consciousness knows the decision, because the subject's subconscious mind has already made a decision which can then be seen on a brain scan. Interestingly, this also shows that even parts of the prefrontal cortex (the seat of the conscious mind) are included in the subconscious mind - further highlighting the power of the subconscious to guide our daily behavior.

"One of the unfortunate things about our education system is that we do not teach students how to avail themselves of their subconscious capabilities." - Bill Lear

The brain, which is two percent of human body weight, uses up vast amounts of the energy of the body - between twenty percent and thirty percent of the glucose and oxygen in the blood. As part of the design of the brain to avoid unnecessary energy consumption, the prefrontal cortex does not use energy to discover the genesis of ideas. However, by not doing this, the conscious mind gets the feeling of being in control. We tend to think that our decisions were made consciously even though the subconscious may have decided well in advance. The subconscious mind does not care what part of the brain gets credit for decisions, it is only interested in ensuring the survival of the body and the propagation of your genes.

As a practical example, if you can reflect on a sport that you have played (or seen a non-professional play) that required quick reflexes, then you will appreciate the fact that the delayed reaction, in returning a two hundred kilometer per hour serve or in hitting a baseball that has curved at the last moment - that 'reaction time' is partly the time for the information to run through your subconscious mind and then get relayed to the prefrontal cortex. Or, if you have gotten very good at any fast paced physical activity then you have experienced that at your best you do not spend much time consciously thinking and analyzing - you just operate in a state of flow, where the conscious mind moves to the background and the subconscious takes over. In this state of subconscious, habit-based, flow, you are able to react to a lightning fast serve, a 140km/h off-cutter, or a rapid curveball. This is where the subconscious is making decisions for you before the 0.2 seconds have elapsed - to respond after the 0.2 second point is to respond consciously and this leads to slowed reflexes. Top athletes, when in a state of flow, are making fine-tuned adjustments and physical responses prior to that 0.2 second time lapse by using the subconscious.

THE MECHANISM OF CHOICE

To continue with the issue of decision making. The thalamus (subconscious) only says to the conscious mind (prefrontal cortex):

"Would you like to do A or B or C or D . . .?"

That is all. The thalamus is a key gatekeeper in the brain that is covertly guiding your every behavior. And the thalamus does not just give you the options 'A or B or C or D,' but simultaneously the thalamus tells you:

> *"If you do A I will give you this wonderful burst of pleasure, called dopamine; if you do B I will give you less dopamine; if you do C or D, I will cut the dopamine levels down so low that you will feel horrible."*

This is what the thalamus calls your 'options.' This is the extent of our 'free will.' Many times options C or D are the options that you logically know are in your best interest - like studying for an exam. Yes, we have the option of consciously choosing C or D, and that is called willpower - but as you well understand, willpower can be exhausting.

So the next time you know in your conscious mind that you should go to the gym, but you really 'feel' like you do not have the motivation to do so . . . you can thank your thalamus for placing 'going to the gym' as option C or D.

Now assuming you drag yourself to the gym at 5am, in spite of the initial drop in your dopamine levels that the thalamus creates - when you arrive at the gym and the workout gets going, you will feel fantastic, and the basal ganglia and hippocampus will take note of this and share that information with the thalamus. The basal ganglia and hippocampus will communicate to the thalamus:

> *"Hey Thalamus, going to the gym was amazing. We enjoyed it and the conscious mind loved getting it done."*

But still, the thalamus will make you have to repeat the process dozens of times before it finally will get to the point where it raises 'going to the gym' to the preferred options of A or B. That is how habits get created.

Someone who has been going to the gym every day for a month will find it much easier to keep doing so, compared to someone trying to self-motivate their way to the gym for the first few weeks. The behavior becomes a habit when the thalamus is finally convinced (based on the statistics taken by the hippocampus and the basal ganglia), that a behavior should become option A or B and not the less preferred options C or D.

All this essentially leads to the neurobiological reality that the thalamus, along with the basal ganglia and hippocampus and also the amygdala, are actively engaged in the manipulation of your thinking. And there is a reason why the four of them do this - for millions of years of evolution, the four of them were the key decision makers in the brain. Your cerebral cortex and then the prefrontal-cortex were the last parts of the brain to have evolved. The cerebral cortex existing in all mammals, and the prefrontal cortex defining us as human. We essentially have our consciousness, wisdom and maybe our very soul, strapped onto a dolphin-like brain. I have never tried to persuade a dolphin to go to the gym at 5am but I'm sure it would take a lot of convincing - unless the dolphin is in the habit of swimming there every morning. This is not to diminish the greatness of the human condition, but merely to acknowledge the magnitude of the challenge we are faced with as we look to alter the trajectory of our life to the greatness that is at our Ideal Parallel World (IPW). We, through the cerebral cortex, add the consciousness, spirit and rationality to this dolphin brained human body avatar. We control our destiny and this body can become a servant of our conscious will, once we learn to communicate fully with it. We are called to bridge the gap between our conscious mind and the subconscious mind.

THE SUBCONSCIOUS AND THE MANY WORLDS

> *The 'wave function' of our lives - the flow. The abundance of possible outcomes in every moment, some more probable than others, what force determines the exact realized outcome when we become aware and conscious of the frame we find ourselves in?*

We discuss here the power each of us has in determining just what possibilities are realized as we collapse the wave function, moment by moment, frame by frame, consciousness by consciousness. The nature of that collapse of the wave function is determined by our self-concept stored in the subconscious mind.

Our subconscious mind is aware of the many worlds occurring simultaneously and chooses the reality we continue to exist in based on our self-concept.

> *"Honestly, sometimes I get really fed up of my subconscious - it's like it's got a mind of its own." - Alexei Sayle*

If we accept what the neuroscience is showing, that the subconscious mind serves to take in massive blasts of information from all your senses and then to focus your conscious mind onto a narrow band of awareness, then, we must see how this links to our objective of Moving Through Parallel Worlds to Achieve our Dreams. In a reality where the many-worlds scenario is unfolding, moment by moment, it is consistent to accept that the subconscious mind would have the responsibility to assess the many branching possibilities and to narrow our focus onto just one or even a handful of possibilities. The subconscious mind is aware of the many worlds.

Our subconscious mind is akin to a quantum supercomputer and can guide us along the path through the many worlds to our Ideal Parallel World (IPW) if we are clear with our intentions. The idea of the subconscious guiding you through a specific route through many worlds may suggest to you either a literal decision making by the subconscious mind of what world can be selected in any given moment, or it may suggest to you that the subconscious may aid you in your conscious decision making, that will then serve your advance to your IPW. I suggest both views are correct but you may choose to embrace whichever feels most comfortable. It is well understood in psychology and we have explained here briefly, that the subconscious mind has the dominant influence on conscious decision making and therefore the pivotal role that the subconscious must play for you to achieve success is inescapable.

To access your subconscious is to access your 'higher self.' To access the fullness of your mind is to access the fullness of your potential. You achieve awakening and approach enlightenment when you access your higher self. This is your whole self – this is about completion of self by accessing all aspects of mind. This higher self has the awareness and control of all unfolding worlds and every existing frame. The subconscious is aware of this many-worlded reality that we are all a part of and can guide you to more effectiveness, greater success and extreme growth within it.

The subconscious mind is actively maneuvering in the background to select the probability that will move you closest to the reality that is most consistent with your deeper self-concept. Out of infinite probabilities and possibilities the subconscious mind has the power to choose that which unfolds before you. So to shift your existence/frames in a desired direction, you must engage in the powerful task of shifting your subconscious. Once the subconscious is aware of what you actually want to achieve it will begin to collapse the wave function in a manner that most serves your interest. The subconscious mind is the guiding force for your entire life. The subconscious mind is shielding or showing numerous potential paths based on what it determines would represent a consistent reality for you. So just as the thalamus only shows you a list of preselected A, B, C or D options, similarly, the subconscious mind is only showing you the option of 'parallel world A' or 'parallel world B' or 'parallel world C,' etc. – or you can call this path A or path B or path C or D.

For example, if you are looking for a new job, and believe that you can excel in any environment, then the subconscious mind will show you a path where you can obtain that outstanding job or the subconscious may guide you on a path that leads to an improvement of your skill and then onto that great job. Conversely if you believe that you cannot adapt to most work environments, then the subconscious will narrow your focus to a handful of jobs that are less ambitious – the subconscious will guide you on a path of mediocrity consistent with your ingrained belief system and prior experiences.

The subconscious mind chooses the parallel reality we shall enter next. Our subconscious chooses where in the field of probability each of our own points of expression will materialize. You see The Secret in this reality, is that the conscious mind is not in the driver's seat. The seat of control lies with the subconscious. To move between parallel worlds we must use our moments of consciousness/awareness to influence the subconscious mind and this influence guides the subconscious to select the particle expressions that most serve your pathway to your Ideal Parallel World.

Later in this pillar we shall discuss how to jack into the subconscious and shift your subconscious beliefs to paradigms that will serve your movement to your Ideal Parallel World (IPW). But first, let us continue to evolve our understanding of how the subconscious is manipulating our view of reality and possibility.

THIS DREAM

When we enter REM sleep and start to dream, we have entered a world that is fully created by the subconscious mind based on the experiences we have had during that day, week and throughout our life. Your subconscious mind is able to deceive you, night after night, to perceive your dreams as real. And then, we must consider that the very same subconscious mind, when you are awake, is using more than ninety-five percent of your brain's processing power. The same subconscious mind that creates the dream world is the busiest part of the mind when you are awake – think about that. What in the world is it doing? The subconscious mind is busy, busy, busy in the background running programs and processes and telling you exactly what to focus on and what to ignore. Is the subconscious mind building the world that the conscious mind is then perceiving? Is our waking reality an illusion?

This is a very valid question. I propose to you, that it is an illusion. I propose to you, that you are dreaming right now. I propose to you, that once you come to fully understand how the illusion works, you will develop the power to alter the world around you. Maybe you have tried lucid dreaming? Now consider how much fun it will be when you crack the code of lucid living. The key part of this crack, is to master the art of planting ideas in the subconscious. To deceive the subconscious with an altered view of reality, very much like it has been deceiving you your entire life. It is a cycle, we can use the conscious to implant ideas into the subconscious -- powerful ideas that once they take hold will move mountains.

> *"And Jesus said to them, Because of your unbelief: for truly I say to you, If you have faith as a grain of mustard seed, you shall say to this mountain, 'Remove hence to yonder place;' and it shall remove; and nothing shall be impossible to you." - Matthew 17:20*

This faith must be in both minds through the pathway of manifesting power: conscious influences subconscious then later, subconscious influences conscious.

Conscious --> Subconscious --> Conscious

Again with our superposition of the literal and metaphoric nature of ideas: You do not have to interpret what I am saying to mean that this hologram of a world is a higher level, denser version of a dream; what I am saying does not mean that you will ever wake up from this 'dream' without having to experience death; but merely accept that the subconscious is so powerful, that to completely connect to it and to develop the ability to alter the beliefs within it, shall give you tremendous control over reality. It would be, for all intents and purposes, as if this world were a higher level dream, being shaped by your will.

For most of our lives we may have thought that our nighttime dreams were unreal, and that our dreams were merely a byproduct of what was really important, that is, our waking reality. The suggestion herein, is that the inverse effect is actually what's occurring -- what we program into our subconscious mind is actively shaping the world when we awake. In a manner of speaking, the important experience is the subconscious experience. The dream mind is actually more powerful than the conscious mind. However, the conscious mind has influence over the dream mind. So by using emotion with the conscious focus on the behaviors, attitudes, habits, lifestyle that one wishes to adopt -- you can create those ideas in your subconscious and subsequently they will manifest into reality.

THREE-WAY MANIPULATION

The subconscious mind continually alters not only your view of the present and the opportunities you see in the future, but even the past gets regularly reshaped by the subconscious. The subconscious mind directly manipulates your memory of all past events. Every time you recall an event the subconscious mind has first made alterations to that memory that will align with your view of the world. This is a well understood phenomenon in psychology called the *misinformation effect*. This is one of the reasons eye-witness testimony in court cases is considered highly unreliable unless corroborated by multiple, disparate eye-witnesses or physical evidence.

> *"Quantum theory means that a physical system does not have a single history, but rather has many histories, each associated with a different probability." – Stephen Hawking*

The opportunity here is that when we, through use of our conscious mind, alter the 'self-concept' that the subconscious has stored about us, we cause the manipulation of our past to occur in a manner that will inspire and drive us.

So when we are on purpose, aligned and on the path to our Ideal Parallel World, the subconscious mind will begin to even alter some memories of the past to ensure that we move to the right path. This may seem like a scary concept, but understand this is ongoing in all circumstances, including right now; what we are doing here by using the Conscious Mind -> Subconscious Mind -> Conscious Mind pathway, is ensuring that when this manipulation inevitably occurs it is in your best interest and aligned with your Ideal Parallel World.

"Events that we think can unfold in only one direction can also unfold in reverse." – Brian Greene, NOVA, 'The Illusion of Time'

HOW OUR SUBCONSCIOUS GETS PROGRAMMED

We program our subconscious mind every day when we feel emotions in response to events. Our emotions imprint ideas onto the subconscious and cause us, in future, to relive experiences that we imprinted. As we go through our day our main purpose must be to place the right programming into our subconscious through attaching high positive emotions to the types of events we want to repeat. All this is to say, simply: feel an exaggerated sense of love and joy and gratitude for the activities, events, circumstances you want to see repeated in your life and feel little to no emotion for the activities and events you do not want repeated. What we focus on, through the lens of emotion, expands. What we feel about, expands.

NINETY-FIVE PERCENT VERSUS FIVE PERCENT

We must realize that ninety-five percent of our behavior is emerging from the subconscious mind, with the remaining five percent from the conscious. So our conscious desires, wishes and hopes are only five percent of what we will do on a daily basis unless we tap into the subconscious mind. There really is no other way. To try to will oneself 24/7 is strenuous and likely to lead to failure, exhaustion, or just inefficient and slow results. That which is done through the subconscious is easy and automatic; it is quite literally the state of flow where our conscious mind stops trying to force everything and we relax and let a much greater force take the driver's seat and move us in the direction we have programmed.

If we did not grow up with the habit of planting only elevating and motivating thoughts into our subconscious mind then the vast majority of the programming in our subconscious mind is what we received from our parents, friends, society and general environment. Some of these may be great programs – maybe your parents raised you to believe you can be anything you aspire to be, and maybe your friends led you to believe that you have the ability to be successful and popular, or maybe your programming is filled with limiting beliefs and self-doubt. Either way, with this book you now possess the power to reprogram your mind, through many methods including daily 'auto-suggestion' and 'self-hypnosis,' and to start to create the automatic behaviors that will serve you well. The big concept to embrace here is that: It is most efficient and stress-free to create habits by implanting the idea in the subconscious mind instead of slogging through the thirty day plus process of creating habits through the will-power of the conscious mind. Ideally you are doing both but noting that the subconscious thoughts account for ninety-five percent of your behavior and the conscious thoughts account for five percent.

Consciously or unconsciously we are always planting ideas in our subconscious through our conscious thoughts, our conscious beliefs, our conscious decisions, our conscious fears, and our conscious affirmations.

All that you think on, focus on or consider in the conscious mind is immediately planted in the subconscious mind. Every conscious thought you have, every moment you spend on an idea is a commitment to be stuck with that idea and with bits of that level of thinking, for the rest of your life. Spending just 10 seconds focusing on a topic that does not serve your interests is to invest your energy along a path that will continue to draw from you and define you. The more you keep your conscious focus on the things you do want, the more the subconscious will lead you on a path to replicating more of the same as you progress in life. Your conscious focuses the subconscious, the subconscious then focuses the conscious in one continuous, self-reinforcing cycle.

It is useful to view yourself, partly, as a computer, and to realize that you can easily upgrade new programs and software into your system. The way to do this is through the subconscious. The subconscious is the non-obvious part of the iceberg below the surface – the conscious is what we deal with above the surface. To attempt to change habits and behaviors through the conscious by motivating oneself during the day can be useful but it is unnecessarily strenuous work. Implant an idea in the subconscious and sit back and watch how your very motivations and drives change to be in line with the implanted idea. Much less work – in fact it would take more work to not do that which you have implanted in your subconscious.

Positive thinking comes naturally to those born of success – so there is a greater likelihood that they will replicate success as they mature. So it is with those born to wealth and the opposite effect with those born to poverty. Our mental programming must constantly be updated and upgraded. This is part of the process of moving to the NIF and just part of normal maintenance in reaction to the nature of our environment.

INCEPTION

You have the responsibility to master the art of implanting ideas into your own subconscious - by so doing you gain the power to change your behavior, you gain the ability to access unlimited information and your power to alter reality grows exponentially. The game of life is being played out in the subconscious - that is where the field of battle is, in the subconscious - the scripts of our entire lives are in the subconscious. To change your life, change your subconscious thoughts.

> "*In a dream, in a vision of the night, when deep sleep falleth upon men, in slumberings upon the bed; then he*

openeth the ears of men, and sealeth their instruction..."

- Job 33 : 15-16

We are all called to engage in an act of idea inception. Inception of an idea, or 'beginning' of an idea, can be understood by considering the 2010 science fiction thriller written and directed by the iconic Christopher Nolan. In the movie, 'Inception,' the protagonist must engage in the challenging task of planting an idea in the subconscious mind of an unsuspecting victim. Similarly, the very essence of strategy in life is to engage in this idea inception – the subconscious mind you are breaking into is your own. It is in your own mind where you must plant ideas of prosperity and abundance. You must plant these ideas consistently until they grow on their own – spread and multiply, like a virus – spreading this contagion of possibility to every neuron far and wide.

Creating a clear image of your Ideal Parallel World is vital and you shall find this image useful once you begin implanting ideas into your subconscious. That image allows you to get on the same emotional and intellectual frequency as the elements of your Ideal Parallel World. By being on the same frequency now as you will be in your Ideal Parallel World you allow your brain to focus on information and ideas that are aligned with the arrival at your IPW.

> *"The conscious mind may be compared to a fountain playing in the sun and falling back into the great subterranean pool of subconscious from which it rises."*

- Sigmund Freud

HOW TO TAP INTO THE SUBCONSCIOUS

The more times you consciously access this link between the conscious and the subconscious mind the stronger will be the connection between the two. So all the activities we are about to outline will take practice, but note that accessing the subconscious mind is a life skill and as such you will be doing it until the day you leave the earth. So embrace these activities and work to employ them all in one form or another. This connection you are strengthening between the two minds is the most important work you can do to guarantee your advance to your Ideal Parallel World.

When we are born we are using closer to one hundred percent of our brain's potential because early in life the conscious assessment of everything is very low. This is part of how children learn as quickly as they do, including multiple languages and to walk and socialize and all the many things they learn in just a few years. As we get beyond ages of five or six we start relying more heavily on the conscious mind. We reduce the connection with the subconscious and start to rely heavily on the structured, formal processes of the conscious mind that are necessary to excel in academia. Much of childhood creativity and imagination is lost as we progressively adapt to the structure and formality of school and eventually work. Re-enhancing that link between the two minds is critical to moving seamlessly to your IPW. In fact, without strengthening this ability, you stand little chance of success.

Next up, are twenty-one methods for increasing your access to the subconscious mind. Each of the methods ultimately brings one to a mental dream state where inception of ideas, paradigms and new behaviors is possible. Each of the twenty-one methods brings you to a point of elevated consciousness, either immediately or more usually through repetition over time. Before you apply these methods, you must remember that your key objective is to align the thinking of the conscious and subconscious with your Ideal Parallel World. Both your conscious and subconscious need to ultimately be in agreement for you to achieve any success. I will list the methods below, and then I will guide you through each individually.

Stream of Consciousness Writing
The Right Brain
Understand The Brain
Create Inception
Nightly Audio
Be Alone
Be Still and Silent
Be Focused
Ask Questions
Pursue A Passion
Create A Dream Log
Listen to Your Intuition
Speak No Evil
Visualize
Create Courage
Chair A Board Meeting
Meditation
Prayer
Self-Hypnosis
Care for Your Brain
Love

STREAM OF CONSCIOUSNESS WRITING

Take some time, at least once a week, to write whatever comes to mind. It is important to clear the conscious mind of thoughts as you engage this. The best way to clear the conscious mind is to close your eyes and focus for twenty seconds on the act of breathing. The more you practice, the more this will tap into the subconscious mind rather than just being a conscious activity. For most people it will begin as conscious and then move to subconscious. Just write freely, with or without a predetermined topic. I recommend creating a daily journal (which you will also use for other activities on here) and leave a section on each page of your journal just for this purpose. After you have been doing stream of consciousness writing for more than a dozen times you shall be able to use this method to subconsciously write on a predetermined topic. Your writing will become a valuable means to access your intuition and a means to produce creative work.

THE RIGHT BRAIN

Engage in any 'right brain' creative pursuit. Art in its many forms will strengthen the link between your conscious and subconscious. Painting, cooking, film-making, drawing, stand-up comedy, poetry, storytelling, gardening and almost an infinity of other creative activities. Just engaging the activities is enough to connect you with the subconscious, because creative thinking strengthens the connection between the more logical parts of the brain and the more creative parts. The subconscious is very strongly linked to creative mental activity.

UNDERSTAND THE BRAIN

Learn more about the brain and the mind in psychology. Read the book "The Brain that Changes Itself" by Norman Doidge M.D. Read books and watch programs about the subconscious mind, but only embrace the concepts that empower you. The subconscious mind is infinite so find material that shows you how to increase the use of that power, and ignore the books that would limit your potential. If you get the opportunity, enroll in a course on Positive Psychology.

CREATE INCEPTION

Inception refers to the beginning of a thing. In your case, it is the beginning of a new idea, paradigm or concept that you are planting in your own subconscious mind. Your mind is most susceptible to inception right before falling asleep and right after you awake in the morning. At these points, you have a brief, approximately sixty second window almost directly into your subconscious. You can connect to your subconscious at these points and engage in what is also called autosuggestion. Think of that moment where you wake up and you are just for a moment still in sort of a dream state, and probably could not answer an analytical question that you could easily handle during the day. It is that moment where if you say an affirmation, your brain does not have the awareness yet to say to you, 'oh that is going to be difficult' or 'no, you are not charismatic, you are boring.' The subconscious mind fully believes what you are telling it at these times. If you hear a negative reply from any affirmation, then you know that your conscious mind is the one doing the replying.

So, every night, just before the Zzzzzs consume you, and every morning just as you awake, read out what it is you intend to accomplish as part of your NIF or IPW. Have a definite date for the goal's accomplishment - but at the same time, you must feel as if you have already achieved this goal, right now.

Step 1: Read goal that has a definite completion date.

> Step 2: Feel the emotion of having already accomplished the goal.

This process immediately alters your brain in a measurable physical manner. This is the work of the brain as you sleep – the brain is strengthening some neurological connections and weakening others. By placing that clear intention of what you intend to create prior to drifting off to sleep you cause your brain to create links between other mental circuits/neural routes and that Ideal Parallel World. This literally creates pathways in the brain to the desired ideal, and these pathways in the brain then translate to the subconscious brain finding pathways in reality, towards that IPW. Let us state this differently:

> *To create a path from where you are in physical reality to your Ideal Parallel World, the brain must first create a mental path from the ideas in your head to the idea of the Ideal Parallel World.*

So the subconscious mind may create a pathway linking a conversation you had a month ago with the new Ideal Parallel World idea you just planted in your subconscious. Then upon awakening you find suddenly you have an insight into what you need to do to move closer to accomplishing that IPW ambition, and then, you realize the link between that past conversation and the present insight; or you may have a dream where you get the idea from a conversation with god himself – which is also true. Metaphysically, a conversation with god himself is always a conversation with you, yourself. That is how the subconscious mind works to help the conscious mind. All that is necessary is that you be specific when you implant ideas, that the ideas be stated positively, and that the idea be laced with emotion so that the subconscious brain knows that it needs to pay attention and that this is not just one of many random thoughts you are having.

NIGHTLY AUDIO

Select the audio that plays as you sleep wisely or do no audio at all (do ensure that the audio you select does not disrupt your actual sleep. e.g. audio with decibel spikes or very engaging audio). The sounds in the background of sleep find a way to seep into our dreams. Remember, even though your conscious brain cannot hear the audio as you sleep your subconscious mind is registering many of these sounds. The subconscious mind pays attention because it may have to wake you if you are in physical danger, and through that process integrates the background sounds into your dreams.

BE ALONE

Every sentence that is spoken near you and every idea expressed by others, is registered in your subconscious. Even if the idea is rejected by your conscious and subconscious, the mind still takes it in and the hippocampus and basal ganglia add it to a database of possibilities. To be alone, is to find a temporary reprise from this bombardment of external thoughts. To be alone is to avail yourself to the possibility of hearing your own internal voice. *"The voice of the world will drown out the voice of God, if you let it."* This does not mean you must become an introvert, unless this is your choice - but merely take time when you can, to connect with yourself and to realign your mental state with that of your IPW. Every interaction with other people, activities and media that are not aligned with your IPW, serves to shift your state in a manner counter to that required to advance to your IPW. Time alone allows you to seize control of your state and to refocus powerfully on your IPW.

". . . the state function changes in a causal manner so long as the system remains isolated . . ." - Hugh Everett, III

BE STILL AND SILENT

At other times, find time away from the noise of your own prefrontal cortex. Moments where you deliberately decide to do nothing and dwell on nothing. To admire a flower, or walk along a trail, or spend time with a pet. Take some quiet time, put your phone on airplane mode, hibernate your PC, throw your television in the garbage and be silent.

> *"Let us be silent, that we may hear the whispers of the gods."- Ralph Waldo Emerson*

BE FOCUSED

Having your conscious mind focused on an idea makes it easier for the subconscious mind to understand what you are trying to accomplish. Your focus should be on your Next Ideal Frame or Ideal Parallel World - or even more specifically your focus can be on something more narrow within your NIF or IPW. Be it a specific goal, a specific behavior or even an interest in a specific person. Narrow, focus is necessary until you master the ability to communicate to the subconscious mind. In just a matter of weeks, as you become more skilled at communicating to the subconscious, you will be able to do more, have more and be more.

To become highly focused for any duration over 90 minutes is to allow the subconscious mind to dominate. Highly focused conscious (prefrontal cortex driven) thinking depletes glucose and oxygen in the blood at a rate much too quickly to be sustainable for more than about 90 minutes, and so the mind either defaults to subconscious processing when extended focus is needed, or the individual becomes exhausted and in need of stimulants or glucose. All this is to say that the mere act of focusing on any one thing for an extended period of time, creates a stronger link between the two parts of the mind. So create a goal in any area and stick to it for days, weeks and months. If your conscious thoughts are all over the place then the subconscious mind will feed you mixed results, uncertainty and more confusion - a vicious cycle. By being focused your subconscious mind can collapse the wave form of the flow of reality in the ways that most serve your advance to your IPW.

ASK QUESTIONS

Ask your subconscious questions and let it find the answers. Instead of searching the internet for something you once knew, ask your subconscious to retrieve the information for you. Give it lots of time at first – might even take a night of sleep. The more you do this the faster the process shall become until it is essentially instantaneous and your memory will be astounding.

PURSUE A PASSION

Do something you love, be it a sport or hobby or passion or person. Doing something you love, gives you such a burst of dopamine in the brain that you feel almost a spiritual energy. This feeling of spiritual connectedness runs through the subconscious to the conscious and therefore strengthens that link. The brain literally creates physical connections between the unconscious neural structures and the prefrontal cortex (conscious mind).

CREATE A DREAM LOG

Create a log of your dreams and think on what they are trying to tell you. This can be a quick scribble into a notepad in the morning, after you awake from sleep. Or even just start the habit of consciously analyzing your dreams as you shower in the morning. The subconscious is always trying to connect with the conscious, just as you, the conscious, are now trying to connect with subconscious. The two minds are meant to be partners. To allow this, as in all relationships, you must listen to the subconscious mind just as you desire it to listen to you. Take a brief minute to quickly jot down some details of your dreams at night - in those you will find insights into challenges you have dealt with consciously, you will find solutions to problems and you will find out the areas where more courage is needed.

The decoding of your dreams takes practice, but once you realize that the images in your dreams are mostly metaphoric and symbolic, then you will become better and better at figuring the meanings. The log allows the subconscious to talk to you, and the more you listen consciously to the subconscious, the more the subconscious will listen to you. If you ask the subconscious to reveal better ways for you two to communicate, it will over time, reveal that information to you. It is like asking a lover *'what can I do to better understand you and to be more connected to you?'* - you may get a surprised or puzzled look at first, and maybe not get an answer right away, but as time moves on, the lover will give you more and more hints, clues and cues if you keep paying attention. To have a log of your dreams is to pay attention.

LISTEN TO YOUR INTUITION

If you have a hunch on something small, go for it. Go with your hunch. You will be more wrong than right at first (you will be more often right if the hunch is in an area where you have already developed expertise. In an area where you are an expert, you may now realize that you already have a link to your subconscious). The mere act of trusting hunches and gut feelings on small matters communicates to your subconscious that you are listening, and that this is an acceptable method of communicating to you. So do this first with small things and gradually you will become proficient at applying the technique to larger pursuits. In the world of work, when you communicate with others, you are still required to give them a logical reason for your decisions - but at the same time, be sure to practice using more and more of your intuition. As you become a master of intuitive thinking, you will learn to go with your intuition to create a hypothesis, and then work backwards to test this hypothesis, and then to establish the logic needed for communicating your conclusions.

SPEAK NO EVIL

Eliminate words like 'can't' and 'impossible' and 'try' from your vocabulary and remember that your words are powerful. Every word you say seeps into your subconscious, so become conscious of your use of language and speak with intent and with caution. In the Bible, God created reality through the 'word,' and you, being made in the image of God, must do the same. Whenever you use the phrase 'I Am,' understand that this is the greatest of spoken expressions - I Am strong, I Am capable, I Am efficient, I Am courageous, I Am on purpose, I Am happy, I Am joyful, I Am love - versus I Am weak, I Am unable, I Am tired, I Am lazy . . . or whatever 'I Am' blasphemes one might say that weakens the self. Every 'I Am' imprints powerfully onto your subconscious and you become that which You Are. In the Bible, when Moses asked God his name, God replied, "I Am that I Am." You, as the image of God, Are that which you say You Are. If you say you are lazy then you will continue to be lazy; if you say you are depressed then you will continue to be depressed; if you say someone else is stupid then almost everything they say subsequently, will sound stupid to you.

Self-deprecating humor shows humility, but often reinforces the negative behavior in your subconscious. So be self-deprecating only about the behaviors you do not really want to change. For example, maybe you are a student and you spend 10 hour days in the library in the middle of February and you are being productive and enjoy doing this, but your friends make fun of you for such - you could be self-deprecating and make jokes about how much time you do spend in the library. But that self-deprecation **is** acceptable because you intend to keep doing that super-nerdy behavior; but you would **not** be self-deprecating if you actually intend to spend **less** time in the library and more time socializing. This does not mean you must not be humble or that you must be humorless, but choose carefully what you say in your humility and in jest. The subconscious mind is completely devoid of a sense of humor - it takes you very literally.

Ensure that the words you say match your intentions. If you intend to be courageous, then in your conscious words ensure that you speak only of courage. Never use qualifiers, e.g. 'I'm brave most of the time except for when I am confronted by a lion;' there should be no 'buts' (unless you are comfortable being afraid of lions - which could be understandable ☺). Even if at first you really are afraid of lions, to speak that fear is to reinforce it. If you must deceive yourself at first and proclaim that 'I Am courageous when confronted by a Lion,' then do so. This initial self-deception will shift to reality the instant the idea gets programmed into your subconscious mind.

VISUALIZE

Continuing with the Lion fear, because of its simplicity, and with disregard for whether or not it is useful to fear Lions (versus respecting them as co-predators - 'Lion' is intentionally capitalized because, you know why). Always visualize the behavior you want to reinforce. Visualize courage against the Lion. You still will, in future, act appropriately when confronted by a Lion, so you would not walk up and whack the Lion on its head, or just enter its turf uninvited – but you would not avoid a safari tour in the Serengeti with friends because of an irrational fear of Lions. Instead, imagine a scenario where you are confronted by the Lion, and you act courageously and appropriately. If you are unable initially to see yourself acting courageously but appropriately against the Lion then imagine someone else in that scenario who is not afraid of Lions.

So in that scenario you could visualize the fearless Steve Irwin-Crocodile Hunter (may he be remembered), confronted by a Lion and imagine him handling the situation calmly. Steve, if encountering a lion in the forest, would probably keep facing the lion, backing away slowly, clutching a weapon in an intimidating fashion, looking large, moving slowly back to the safety of the safari van and also having a clear plan for what he would do if the lion were in fact to charge before then. Contrast this with the approach of someone afraid of lions who might just turn and run for the van and get pounced on in five seconds. Once you can visualize someone else doing that which you fear to do, you can then just reimagine the situation and cast yourself in their place. So, when the Lions of your life emerge, do you act calmly and strategically or do you get emotional and then eaten alive? The trick is to always visualize yourself doing the former, and never the latter. Visualization is a technique employed by many top athletes who, every day, during training, imagine themselves winning. The most decorated Olympian of all time, swimmer Michael Phelps, is a famous example:

> *"He's the best I've ever seen and maybe the best ever in terms of visualization. He will see exactly the perfect race. He will see it like he is in the stands and he'll see it like he's in the water."– Michael's coach, Bob Bowman*

You want to visualize frequently. You must get to the point where on your command you can create a world around you with elements entirely conceived in your mind. You should practice imagining a stack of $100 bills on your desk that does not exist, or even something as small as a stapler on your desk that is **not** really there. Be careful here – this is not about embracing madness, this is very much a controlled and conscious behavior. Just as whilst driving you can choose to ignore the insult of a stranger in another car who you will never see again, you can choose to imagine something out of your life that is in your life and imagine something in your life that is out of your life.

If you desire to be confident you must visualize and imagine yourself as confident. If you intend to be the greatest scholar in the history of academia, then you must imagine that right now, at this moment and not at a future time, that you are now this greatest academic scholar. You must be now what you desire and this requires imagination and then visualization.

The subconscious mind stores every single experience that you have and is constantly working to create a present that matches the past. The goal of the subconscious is to create your present reality based on the experiences you have had in the past, but the subconscious mind places more weight on recent experiences you have had. Fortunately, the subconscious mind cannot tell the difference between that which you visualize and that which you actually experience, so you can regularly visualize the life you would like to have and thereby cause this to become part of the subconscious mind's perception of who you really are - essentially the subconscious mind thinks that everything you visualize are events that actually happened and it becomes your duty to continue to mislead it in that regard. This is why imagination is so important. Imagination allows you to create a future that excels your past.

Use as much detail as you can when visualizing. If you are trying to manifest a dream house, you should go beyond just seeing a picture of the house; you should advance to visualizing in 3D, where you can actually walk through the house and make contact with the items therein and have experiences in the house.

How To Visualize

1. Sit down comfortably in a place where you will not be disturbed for ten minutes, and set an alarm to alert you when your time is up (a song, as an alarm, is best).
2. Decide specifically what success or achievement you want to visualize.
3. If available, view a picture of the environment where this achievement takes place (for example, a picture of the Serengeti, or a picture of your workplace).
4. Close your eyes.
5. Slow your breathing by taking very deep breaths. Consciously take control of and monitor your breathing.
6. Imagine your body relaxing, one major body part at a time from top to bottom: Head, shoulders, arms, legs then feet.
7. Keep breathing deeply. Imagine each body part more relaxed than before.
8. Imagine yourself in the environment with as much detail as you can create (as with all things, practice makes perfect - you can even visualize yourself visualizing to get better at this ☺).
9. Awake when your music starts (or timer goes off). Acknowledge your awesomeness.

With practice your visualizations will become more real and even you conscious mind will be momentarily startled when you awake. When you get to that point where even your conscious mind gets sucked into the illusion, your focus and motivation will increase exponentially. This is the point where the two parts of your mind have aligned, albeit temporarily. At this point you start to operate with a level of grace, confidence and unflappability that leads you to do exactly what you need to do, exactly when you need to do it. It is a very powerful state of continuous flow, that if maintained, sheds light on your path and enlightens your soul.

CREATE COURAGE

Very little of this works if you have major fears dominating your life. Confront your fears, because fear clouds your link to the subconscious. The wise words of Master Yoda in the Star Wars films should always be heeded. In the Star Wars saga, George Lucas identified fear as the very path to all evil and the misuse of power, which he called the Dark Side.

Fear causes the amygdala to shift the brain into fight or flight mode, and releases surges of adrenaline and cortisol. All this is very useful if you are in imminent danger but is disastrous to the body after more than a brief period. If you have major Lions in your mind that pop up throughout the day, your link to your subconscious is constantly being shut down by the amygdala. Eliminate fear by being open to possibility. See yourself overcoming the worst and continuing to thrive.

CHAIR A BOARD MEETING

You can create a list of six people (or fewer), alive or dead, fictional characters or historical people. Your list could include Gandalf The White, Abraham Lincoln, Jesus Christ, The Buddha, Mother Teresa, Ronald Reagan, Michelle Obama, Justin Timberlake, Condoleezza Rice, Warren Buffet, Martin Luther King Jr., Abraham Hicks, Jack Welch, Napoleon Hill, or any person who you would define as successful and whose advice would help you advance to your Ideal Parallel World. Learn as much as you can about each of these persons and what their personalities and thoughts were like. Then, once every week, at the end of one of your meditation sessions, you are to imagine you are at the head of a table with your list of six as invited guests. With your eyes closed, and still in a meditative state, imagine that you are channeling these six people to join you. They are invited for the purpose of giving you advice on how to progress to your NIF and IPW. This activity will feel entirely fake at first, but each time you do it, it shall start to become more dynamic and realistic. By your fifth session the characters start to take on a life of their own that is so compelling that you will feel like you are actually channeling them - and maybe you are. As chair of the board you may ask them questions, and their advice will achieve for you, exactly what their advice would have achieved for you.

MEDITATION

Meditation is one of the most important activities you can engage in for connecting to the subconscious. Essentially, every activity that we have listed here is a form of meditation. Connecting to the subconscious is all about quieting the conscious mind and listening to that inner voice, or in a manner of speaking - listening to the higher self. Meditation is the most condensed or natural form of connecting to the subconscious because it is all about achieving a state of silence. There is no need for affirmations or visualizations in the meditative state - just a need for peace, and bliss and silence. Use meditation as part of your daily routine. Meditating twice daily for any amount of time - one minute or ten minutes or fifteen minutes, increasingly brings one to a state of centeredness, a state of power and to a state of connectivity to the infinite. Meditation has been shown in an abundance of studies to impact every aspect of health, from cardiovascular, to immune, to psychological wellbeing and even recently to the lengthening of cellular telomeres - which means a slowing (technically a reversal) of the effects of the aging process.[28]

The more you practice meditation, the faster you will be able to enter a meditative state. I recommend 1 minute meditation throughout the day, when you are most busy. If you are too busy to meditate, then you definitely need to meditate.

How To Meditate

"Calmness is the ideal state in which we should receive all life's experiences," - Paramahansa Yogananda

The process of meditation is similar to visualization and self-hypnosis. The key difference is that with meditation the intention is to create total silence, that is, no affirmations or visualizations, although you may choose to end your meditative sessions with affirmations or visualizations to maximize the use of that meditative state.

1. Sit down comfortably in a place where you will not be disturbed for 10 minutes (set music/alarm to alert you when your time is up).
2. Close your eyes.
3. Slow your breathing by taking very deep breaths. Consciously take control of and monitor your breathing.
4. Imagine your body relaxing, one major body part at a time from top to bottom: Head, shoulders, arms, legs then feet.
5. Keep breathing deeply. Imagine each body part more relaxed than before.
6. Maintain focus on your breath. If thoughts pop into your mind, just say hello to them and then refocus on your breath.
7. Awake when music/timer goes off.

PRAYER

To pray is to be in a state of openness, allowance and peace. Prayer creates a state of mind very similar to meditation. Prayer focuses the mind. It creates a connection to a powerful source of energy that I need not define.

> *"I like the silent church before the service begins, better than any preaching."- Ralph Waldo Emerson*

SELF-HYPNOSIS

Learn to hypnotize yourself and create your own affirmations. I have included below some general steps for self-hypnosis to achieve your NIF. They can be applied similarly to your IPW and for other needs.

I personally recommend that you download self-hypnosis audio online. There are free versions available on YouTube® but if you are looking to purchase a great self-hypnosis series, I recommend: Lyndall Briggs, Live Life Powerfully: Seven Clinically Proven Guided Visualizations Geared To Turbocharge your Life: Available on Audible®.

How To Self-Hypnotize To Achieve NIF

1. Lay down or sit down comfortably in a place where you will not be disturbed for fifteen minutes.
2. Close your eyes.

3. Slow your breathing by taking very deep breaths. Consciously take control of and monitor your breathing.
4. Imagine your body relaxing, one major body part at a time, from top to bottom: Head, shoulders, arms, legs then feet.
5. Keep breathing deeply; imagine each body part more relaxed than before.
6. Say to yourself that this session of self-hypnosis will be effective – you are intending that self-hypnosis will work now and every time, just like you intended for this book to work.
7. Imagine a great success at your NIF and the congratulations from your peers at your NIF.
8. Imagine yourself engaging in all the right behaviors to achieve your NIF.
9. Feel the emotions of having already achieved your NIF.
10.Take more deep breaths and imagine yourself meditating and engaging in hypnosis every day.
11. Engage the inception of positive affirmations by speaking now to your subconscious.
12. Take a few more deep breaths, arise. Feel awesome.

CARE FOR YOUR BRAIN

Exercise and nutrition are critically important for your brain. The brain functions mainly on oxygen and glucose from the blood, so improved cardiovascular health translates to a brain that is able to perform more efficiently and effectively. Exercise and proper nutrition lead to an overall well-being that allows one to exist in a higher energy state and allows the brain to communicate more efficiently within itself. This translates to clearer thinking and a greater overall awareness. I strongly recommend that you include behaviors related to improved exercise and nutrition as part of your self-hypnosis.

LOVE

To plant the image and behaviors associated with your NIF and IPW in your subconscious throughout the day, go ahead and use the compelling emotion of love - just as you would in a romantic scenario. Really allow the ambition and necessary behaviors to consume your waking thoughts and surround them with the emotion of love. Love is a powerful form of hypnosis. You must develop the type of focused romantic love that energizes and releases chemicals in the brain that multiplies your physical and spiritual energies. This type of passionate love for that which you desire, and passionate love for the behaviors that lead to it will allow you much easier access to the subconscious mind than you would ordinarily enjoy.

As we learned, the subconscious mind can be viewed as the 'dream-state' mind and I am sure if you think on some of your favorite strong attractions and loves in your life there was plenty of day-dreaming involved. Think on that term, 'day-dreaming,' for it suggests a direct access to the subconscious mind even whilst wide awake. That type of access is powerful.

You must create love for your NIF and through love the idea will get firmly planted into the subconscious. Once the idea is implanted into the subconscious, your conscious mind will start to receive clear directives as to the behaviors you must engage to make that NIF a reality. You must then attach your love to those behaviors. So in romantic love, you would love the process of going on a date or love buying flowers or love complimenting the person of your desire or love being there to support this person - these are all behaviors that will serve to win over the subject of your desire. The process is the same when you attach love to the NIF; you must not just love the NIF but love the behaviors that will bring you closer to the NIF. For example for achieving many common NIFs you will need to exercise daily and would have to learn therefore to love exercise, and to love proper nutrition. For another NIF you might have to learn to love academic study, or love a specific subject or love a certain type of mental training.

With all NIFs you will have to learn to love meditation, or self-hypnosis and love positive thinking and to love courage and even to love, love. Using love to focus is important and the focus is largely about giving clear impressions to the subconscious mind.

GUIDELINES FOR PLANTING IDEAS IN THE SUBCONSCIOUS

1. Always state ideas in the positive:

I am always on time for business meetings. ☺

~~I am never late for business meetings~~

2. Always state ideas as past or present tense (any tense other than future tense, even if a future date is used).

I make 1 million dollars selling real estate. ☺

I made 1 million dollars selling real estate. ☺

I am making one million dollars selling real estate. ☺

Between July 2020 (future date) and December 2021, I made 1 million dollars selling real estate. ☺

Between July 2020 (future date) and December 2021, ~~I will make 1 million dollars selling real estate.~~

3. Use as much emotion as possible when planting ideas or visualizing. The subconscious mind pays attention to emotion. The subconscious mind influences your conscious mind by altering your emotions. This is the human decision making process in a nutshell: we make decisions with our emotions and then we justify them with logic. Therefore, if you want to seize control of your behaviors you must use the reverse of this effect: you must choose your behaviors with logic and then justify the choice to the subconscious mind by using emotion.

"The more intensely we feel about an idea or a goal, the more assuredly the idea, buried deep in our subconscious, will direct us along the path to its fulfillment." - Earl Nightingale

Indeed, every thought you have is planted into the subconscious mind. The greater the emotion attached to the thought the more dominant it becomes in your subconscious and the more results you will experience in your life that are consistent with that thought.

4. Keep strengthening the connection between the conscious and the subconscious. Some people reduce or stop their meditation and self-hypnosis once they start accomplishing some key goals and feel like they have a great momentum going. Often this is done because they fail to realize that this connection between the conscious and subconscious is the true source of their success. They erroneously give credit for their progress to the conscious mind – the ego. This sets the individual up for a plateau of performance or worse, a regression back to mediocrity. It is important that you continue with the activities that impress on your subconscious mind for the duration of your life. To ensure this, I strongly recommend that two of the ideas you plant in your subconscious mind are:

i. I engage in meditation/hypnosis twice or more daily, forever

ii. I can access my subconscious mind through peaceful moments when I am awake.

These two additional meditations will serve to keep you on track and are essentially a cycle. That is, so long as all your meditations are working, then these two meditations will also work and reinforce the entire process.

THE IMPORTANCE OF THE SUBCONSCIOUS MIND

We have outlined the key methods for tapping into the subconscious mind. To fully master this pillar we need to continue to understand the importance, role and functioning of the subconscious.

AUTOMATION OF REASONING

The human brain is automating as many processes as it can - shifting them to the automatic, which is the subconscious. Therefore as we progress through life we become less conscious of the most constant elements of our behavior. Our very existence becomes automated and our ability to change our self through solely the conscious mind faces greater restrictions. This is another reason why we must focus on the subconscious as the pathway to behavioral change and so too the pathway to our Next Ideal Frame or Ideal Parallel World.

THE SUBCONSCIOUS AND NEXT IDEAL FRAMES

> *"Finally, brethren, whatsoever things are true, whatsoever things are honest, whatsoever things are just, whatsoever things are pure, whatsoever things are lovely, whatsoever things are of good report; if there be any virtue, and if there be any praise, think on these things." - Philippians 4:8, King James Version (KJV)*

By focusing on the very specific Next Ideal Frame (NIF) we want to achieve, the subconscious mind draws in information that is relevant to this NIF. Your subconscious mind then feeds the 'frequency' consistent with your NIF to the conscious mind thereby strengthening your decision making and intuition en route to your NIF. By 'frequency,' I mean information that is consistent with a train of thought. Just as by choosing a radio station you get only the broadcasts on that frequency, so too by focusing the subconscious on a specific NIF the subconscious then provides your conscious mind only the information related to that NIF. So our subconscious mind focuses us only on specific frequencies of information from our environment, based on how clearly we programmed our NIF into our subconscious. The focusing of our conscious mind on specific frequencies aligned with our NIF helps determine the nature of the collapse of the quantum wave function of possibility.

THE PATHWAY TO GENIUS

Creating a link between the conscious and subconscious is a pathway to using the fullest capabilities of the mind. The genius is he/she who creates a strong connection between the subconscious mind and the conscious mind. The mad man is he who allows his subconscious mind to dominate all his thinking. This is why genius is so often typified by some degree of madness - the approach to genius is in fact a flirtation with insanity.

> *"There was never a genius without a tincture of madness." - Aristotle*

THE FIELD AND THE SUPERCONSCIOUS

The subconscious mind has a connection to a massive 'field' of knowledge. Some teachers have called this the 'superconscious' and others have called it 'the field.' This source of knowledge exists either through a metaphysical connection, or may simply be the result of the fact that the brain stores every memory that we have had in our lifetime. This includes every single detail of every environment we have been in, and every conversation that has occurred in ear shot, and any data that we have been exposed to – none of this needs to have been in our conscious awareness as the subconscious mind has the biological responsibility of letting the conscious mind know when to pay attention. So once we realize that the subconscious has this massive store of data, then it becomes obvious that it can provide us with astounding and seemingly infinite wisdom when we learn to connect to it.

> *"My brain is only a receiver; in the universe there is a core from which we obtain knowledge, strength, inspiration. I have not penetrated into the secrets of this core, but I know that it exists." – Nikola Tesla*

Use of the subconscious mind to connect to the broader field – what Napoleon Hill calls 'the ether' is also part of being in the state of what psychologists call flow. In quantum physics we can think of being in flow as being connected to that infinite wave of superpositioned possibilities.

LIFE SHOULD BE EASY

Ultimately, all the work that you do should feel easy and fun and free flowing. Hard work means that you are working against your subconscious instead of with your subconscious. Hard work means that there is this dissonance between the two parts of the mind. It means that you still have much work to do on strengthening your connection with your subconscious mind. Through this strengthened connection you will be more able to shift your self-concept to one that is aligned with reaching your Ideal Parallel World.

We feel discomfort, cognitive dissonance and generally ill at ease when our conscious and subconscious are focused on different ambitions. When the programs we would like to execute consciously are different than the programs we have in our subconscious we feel the pangs of an addict's life. We feel out of control when we consciously have no control over the subconscious. Some of us lucky few may have great programming from childhood and are living lives of abundance; but if you have not been so lucky and picked up numerous bits of ineffective programming from family, friends, neighbors and television, then there is the real possibility to write your subconscious programs anew and begin the journey of a focused and aligned life.

Indeed, there is this real potential for conflict between the conscious and the subconscious if we do not adjust our subconscious to our conscious ambitions. This is the mechanism of self-sabotage when an individual has a mental programming that says life is supposed to be difficult instead of a joyful adventure, that we are supposed to work hard for money instead of having money work for us, that romance only happens in the movies instead of having romance be part of everything that we do, that our skin color is a limitation instead of being either irrelevant or a bonus. What happens is the subconscious mind is looking to create a reality that matches its programing and if we are consciously looking to retire at age 40 and our subconscious mind thinks we should be a slave till the grave, then self-sabotage and inexplicable failures are the result. In sports we see teams 'choke,' those are all the results of dissonance and discordance between the conscious and the subconscious. When our conscious mind believes we can win, and our subconscious mind believes we can win, we never choke, we never stumble – this does not mean we always win, but it does mean that we wield a force of focus so powerful that it would take quite an obstacle to stop our advancement.

Every time you are engaging in a behavior other than one that serves your movement to your IPW and every time you engage in a behavior that is other than what you need to be doing or a behavior that you know is self-sabotaging or limiting, this is a sign of your subconscious brain at work. You may feel like what you are doing is just exactly what you feel like doing, or that you need a break, or that you are too tired and lack the willpower right now – but understand the reason you need willpower at all is because you are fighting against the desires of that ninety-five percent of your brain that wants to do this **other** 'fun' or self-sabotaging activity. I am not saying you cut fun from your life, what I am saying is that the primary reason you perceive an activity as fun is because your subconscious mind says it is fun. The whole concept of fun is subjective to begin with. All that you consider fun is simply the release of the neurotransmitter dopamine through the neural pathways in the brain that control the behavior you are engaging. Your brain has these neural pathways it creates for every possible behavior.

These pathways are connections of neurons and when the neurons fire together along a specific route it triggers one specific behavior. Introduce a flow of dopamine to the prefrontal cortex whilst this is happening and you shall perceive the activity as fun. If you think of the most boring task imaginable and introduce a flow of dopamine whilst doing so, then that boring task will be the most pleasurable thing you have ever done. This is partly how drugs for ADHD work in creating the sensation of pleasure for a kid who might otherwise be bored out of their mind sitting in a classroom – they pop the amphetamines or Ritalin® and all of a sudden even boring classes keep them in rapt attention. But the process of dopamine release is affected by perceptions of the subconscious mind. The same subconscious mind that tells you that food is delicious can also tell you that going to the gym, or studying all night, or saving money, or any other great habit, is super fun!

Through planting the right habits in the subconscious mind you will come to redefine fun as 'that which serves the interest of your growth and movement to the NIF and IPW.' This does not mean you become a recluse and stay away from all social and delightfully hedonistic pursuits. This merely means that, for example, when you make a decision to spend the night having drinks with friends, that you make that decision, you go out and have a great time knowing that this activity lifts your spirits, strengthens your social skills, strengthens your network of support, creates something wonderful for you to feel grateful for, or countless other positive benefits of doing traditional fun things. But, what I am saying is that you would not go out drinking with your friends if you know it is just a bad idea, will make you feel terrible in the morning and cause you to not accomplish some goal you are working towards; and you would not go out with those friends if you believe they are having a toxic impact on your life. Your life then becomes one where you are engaging in the behaviors that you choose to engage in, and you are not just a slave to prior programming.

Your subconscious mind realizes that playing video games for a couple hours with friends on a Saturday night might be quite fun and inspiring but if your subconscious mind knows what your Next Ideal Frame looks like, after the third hour of Final Fantasy it will likely cause you to crave a more intellectually stimulating activity. This will not be a painful process, this will actually be the most fun you have had in your entire life. You create a life where the conscious and the subconscious are speaking coherently to each other and are working as a team - that I can assure you, creates a feeling of passion, focus, flow, intention and this conscious -> subconscious path is the only path to your IPW.

CHAPTER CONCLUSION

The best part of this whole process of working with the subconscious mind is that once we become adept at communicating directly to our subconscious mind the whole concept of effort withers away. Connecting to the subconscious eliminates the concept of being lazy or lacking drive or energy - this is because, human beings are always doing something, even if that thing is sleeping or watching television or playing video games. So our bodies have the ability to 'do something' for the entire duration of our lives, it is just that often times when we feel unenergetic, the 'something' we feel to do is different than the 'something' we need to do. So we may consciously want to study for a final, but our body really just wants to play Assassin's Creed or watch the latest MTV show. Understand that whenever you are doing something that your conscious mind thinks you should not be doing, that is, whenever you find yourself being 'lazy' or unmotivated, it is the subconscious mind that has made a choice that is against the conscious mind. So by reprograming the subconscious to have the same motivations as the conscious the whole process of motivation becomes irrelevant and we have a life where we just naturally do exactly what our conscious mind rationally thinks we should be doing (this includes recreation). Interestingly, even when the two minds are aligned, there will be times when you are driven to do something that seems counterproductive - you shall perceive this as a 'gut feeling,' where the subconscious mind has made an advanced calculation of numerous variables, and is driving you to do that counterproductive thing. That scenario often plays out in our best interest, but even the subconscious mind is making probabilistic calculations and will on occasion steer you to an unsuccessful endeavor – but this will always be a risk worth taking. For example, if the subconscious mind determines you have a ninety percent chance of meeting up with someone at a specific location at a specific time, and this is a person who can help you achieve a goal that you have programmed into the subconscious, then guess what? The subconscious may cause you to take a path that leads you to that individual, or cause you to catch an earlier or later train so you serendipitously have this meeting. Yes, a ninety percent chance is a good gamble but it does mean that ten percent of the time you would have merely been sent on a failed errand by the subconscious. Sometimes, the subconscious mind may take a circuitous path to your IPW – this is necessary. Much of what you must learn in order to succeed is not obvious or conscious. On balance, trusting your gut, instincts, intuition, all of which are feelings from the subconscious, is sound action.

I must state that our connection as human beings to that which is infinite, however you describe this infinite force or being, is along that pathway: Conscious -> Subconscious -> Infinite. This is why many of the actions I discussed for strengthening this link between you and the subconscious gives one a feeling of spiritual connectedness and purpose in life. It gives you a connection to God, or god or superconscious, or the cosmos or whatever concept you choose. This pathway is a link to infinite power. It connects you with the flow and provides you with the unfathomable power to collapse the wave function of reality in the manner which you see fit. The speed with which you advance to your Ideal Parallel World is in direct proportion to your ability to build this bridge.

As you strengthen the **Conscious --> Subconscious --> Infinite** pathway, the reverse pathway also becomes stronger. Yes, the reverse pathway. As you strengthen the **Conscious --> Subconscious --> Infinite** pathway you will be able to hear the very voice of god, or God, or the cosmos. You will receive guidance from **Infinite --> Subconscious --> Conscious**. It is that moment where you look at reality that may have once knocked you down and suddenly you can see the very photons of the hologram, and you fear nothing, not even death, and know that you can remake the world to match your ideals. That is the prize in store for mastering the concepts in this pillar. This is why all great spiritual teachers throughout history engaged in meditation and prayer. This is one of the reasons prayer works. There is a very real connection to a source of power that emerges from the building of this connection. To achieve this powerful connection is the next leap in human evolution, and you now have a head start.

This illusion that is our physical reality. An illusion that becomes most apparent with the measurement problem in quantum physics, and part of the reason that all matter is comprised of more than 99.9% empty space, and the dominant elements of matter are what can only be described as electrical impulses, protons and electrons. The entire universe is of this form -- it is an illusion created before your very eyes; but the delightful, fitting and eternal irony of this illusion, of this ruse, if you will, is that, you are the creator. You are the player, you are the source, you are the god. To engage the power of the subconscious is to connect with your higher **self** – to connect with the Infinite – to connect with the Source.

PILLAR 2

IPW THINKING

"We are what we think. All that we are arises with our thoughts. With our thoughts, we make our world." – The Buddha

IPW Thinking, is the second, powerful pillar for Moving Through Parallel Worlds To Achieve Your Dreams. IPW Thinking is in part, a paradigm. It is a manner of thinking that draws your Ideal Parallel World to you, and draws you to your Ideal Parallel World. In that sense, it is like a gravitational force. Yes, IPW Thinking is like gravity. IPW Thinking is a force of alignment – it is an unseen force that keeps you aligned with your IPW. It is the energy that keeps you on the right path, both consciously and subconsciously. To use IPW Thinking, is to use the power of the force - the force that lies between you and your Ideal Parallel World.

Every time you get your thoughts into alignment with your Ideal Parallel World, you are engaging in IPW Thinking. When you are in alignment with your IPW it means that you are using IPW Thinking, otherwise you would fall out of alignment.

To engage IPW Thinking is to be in a state of flow, as to use IPW Thinking is to be in the flow. To use IPW Thinking is not just to be in the flow, but specifically it is to be in the flow and in the specific fast moving current that lies between you and your Ideal Parallel World.

IPW Thinking is a state of connectedness to your IPW. To be engaged in IPW Thinking is to embrace a mental state of allowance and knowingness. To engage IPW Thinking is to embrace the inevitability of success – engaging IPW Thinking is knowing the end and being the end. To be immersed in IPW Thinking is to say, 'it is inevitable – success, is inevitable.' This attractive force I introduce to you, is an idea in superposition: it is literal and it is a metaphor.

IPW Thinking is defined by the thought processes and paradigms that will allow you to move to your Ideal Parallel World (IPW) in fewer frames, and with greater ease. The consistency with which IPW Thinking is used will determine your rate of progress to your IPW. You will use both your conscious and subconscious to engage in IPW Thinking. IPW Thinking will become your dominant mental paradigm.

THE THREE PARTS OF IPW THINKING

1. See the Future & Be the Future
2. Positive Emotions
3. Thoughts Creating Reality

We next take a closer look at the 3 parts of IPW Thinking by assessing why and how they work powerfully to get you to your Ideal Parallel World. We shall then discuss IPW Thinking more generally, and also practically.

IPW Thinking Key #1: See the Future – Be the Future

VIEWING THE WORLD THROUGH THE EYES OF THE PERSON YOU WILL BE UPON ARRIVING AT YOUR IDEAL PARALLEL WORLD

This process includes both seeing the present world through the eyes of your future self as well as visualizing the environment of your future world. This is: See the Future – Be The Future. This is assuming the emotions now of the great person you see your future self as.

This is living in the state of *Future-Present*. We all have the choice to either exist in the *Past-Present* or in the *Future-Present*. To exist in the Future-Present is to feel now, as you would feel living in your Ideal Parallel World and to see yourself now as the person you would be in your IPW. To live in the Past-Present, one simply continues on with their current life trajectory. Continuing to live in a reality where each past moment imprints onto the mind the past as a model for the future. The images of the past must never be allowed to limit your future. If left unchecked your mental past will shackle your mind. Always conjure up new images and greater imaginings. At the Ideal Parallel World you are on purpose and the journey there requires you to feel now as you will feel then – you must be now, on purpose. This is an adventure now; this is a quest now.

To focus on your IPW is to achieve mental clarity. To feel now, as you would feel at your Ideal Parallel World creates both conscious and subconscious alignment. It is upon alignment, that your subconscious then becomes able to light the pathway to your IPW.

THE SIGHT

You must envision your IPW as already existing. It must be entirely real to you – you must see it with your waking eyes and feel the feeling of it existing at this very moment. This feeling is achievable, though it may feel new and different, like you are wandering into some newfound madness; or it may feel old and familiar like the imaginary worlds of your childhood. Remember, the pathway to genius can be frightening because it is a journey to the unknown, it is a flirtation with madness. What is madness but the ability to see the world as differently than anyone else; and if this defines madness, then how would you define someone like Richard Branson who sees business opportunities where others see nothing, or Martin Luther King Jr. who had a dream of a more united America, when others could not. Genius emerges from seeing that which is not commonly seen. The genius reaches for something greater, something that is outside of common awareness. The genius is Roger Banister who saw the potential to run a mile in under 4 minutes when all others thought such was physically impossible. The genius sees further and as such his worldview is solitary.

"Talent hits a target no one else can hit; Genius hits a target no one else can see." - Arthur Schopenhauer

To achieve greatness and to progress over to your Ideal Parallel World requires vision. You both see that which is unseen and feel that which is unfelt. You see the future and you be the future. See the Ideal Parallel World and engage the emotions of being at the Ideal Parallel World. Are you confident at the Ideal Parallel World? Then be confident now. Are you fearless and relentless at the Ideal Parallel World? Then be fearless and relentless now. Are You happy, passionate, blissful, outspoken, charismatic? Be all this now. To feel these emotions now is to tap into the energy of your future self.

You created the image of you IPW to allow you to visualize the future as the now. When you visualize:

See yourself engaging your future occupation and imagine the full extent of the experience, with a focus on feeling the emotion of being there right now.

See yourself walking through your future mansion or ideal home and imagine the full extent of the experience, with a focus on feeling the emotion of being there right now.

See yourself as interacting with your circle of friends and family, as well as imagine the new friends you will acquire and imagine the full extent of the experience, with a focus on feeling the emotion of being there right now.

See yourself handling the future wealth, assets and resources and imagine the full extent of the experience with a focus on feeling the emotion of being there right now.

Brand the clear vision of your Ideal Parallel World onto your conscious and use the tools from Pillar I to plant the vision into your subconscious. Your Ideal Parallel World is your North Star – it is the world you are moving towards by seeing it as real today and feeling it as real today. Bring to the present the emotion, the energy and the feeling of power, peace and engagement that you will have at your Ideal Parallel World. You must: "*Begin to be now, what you will be hereafter.*"

Basically, for every item that you included in your image of the IPW, you should imagine the full extent of achieving it and interacting with it, with a focus on feeling the emotion of being there right now.

WHY NOT SEE THE IDEAL PARALLEL WORLD AS FUTURE?

When you visualize you must **never** see the IPW as in the future because if you do so, then it will always be in the future. Your subconscious mind will embrace the idea that this is all stuff that will happen in the future and will then create a life of 'will happen in the future,' forever. You must visualize the items in your IPW as occurring right now. All this is IPW Thinking.

REAL EMOTION

IPW Thinking works when you achieve real emotions - this is because the subconscious mind is forced to pay attention whenever you feel real emotion. The subconscious mind has the biological responsibility of keeping you alive and propagating your genes, so when the conscious mind feels powerfully about something the subconscious mind takes notice. Emotions are the language of the subconscious mind - as we described earlier, its control over your emotions is what gives your subconscious so much control over your behavior. You must visualize the events in your IPW with real and powerful emotion.

IPW Thinking Key #2: Positive Emotions

EMPLOYING A DOMINANCE OF POSITIVE EMOTIONS (INCLUDING GRATITUDE AND LOVE) OVER NEGATIVE EMOTIONS

The positive emotions, in particular gratitude and love, are energetic fuels that give you the power to move through parallel worlds. This is an analogy, but the central idea being that through the impact of various neurotransmitters in the brain, the positive emotions that you feel give you additional bursts of energy to function, exactly when you need it. From an evolutionary point of view, although partly tautological, that is the very reason the positive emotions feel good – the emotions feel good because **the emotions** are good. So to move to your IPW requires that you feel good now and not wait until you have arrived. You are not on a path to feeling good, feeling good is the path.

Positive Emotions

Joy, Love, Gratitude, Bliss, Peace, Empowerment, Freedom, Confidence, Passion, Enthusiasm, Belief, Engagement, Eagerness, Optimism, Hope, Humility and many of their synonyms.

Negative Emotions – The Dark Side
Fear, Doubt, Disbelief, Worry, Anger, Jealousy, Envy, Greed, Lust, Revenge and many of their synonyms.

MOVING UP THE ENERGY SCALE

The more you carry out your day with an intentional focus on feeling positive emotions above all else, then the greater will be your energy and ability to move rapidly towards your Ideal Parallel World. That palpable energy that fills the room when an über-successful person walks in, can be charted. The higher one moves up the scale of positive emotions, the great the power one possesses.

1. Enlightenment
2. Bliss
3. Joy
4. Unconditional Love & Gratitude
5. Happiness & Fun
6. Allowance, Acceptance & Openness
7. Fairness & Reason
8. Ambition & Passion
9. Courage
10. Faith
11. Ego

12. Anger
13. Lust & Desire
14. Fear
15. Sadness
16. Depression
17. Guilt
18. Dishonor

To engage in IPW Thinking is to wake up every morning focused on moving oneself up that energy scale. Set your intention to spend the day as high up on the scale as you can achieve. If you fall down the scale, work your way back up. All this is IPW Thinking.

> *"We are shaped by our thoughts; we become what we think. When the mind is pure, joy follows like a shadow that never leaves." – The Buddha*

SHOULD WE EVER FEEL NEGATIVE EMOTIONS?

It is not that we must never feel the negative emotions. For all emotions have validity and relevance. Let us say we feel weak and perceive that a system is oppressive to us, then moving up the energy scale to anger would be more productive than staying in depression. However, as we will explain in this chapter, to stay in a negative emotional state for many days at a time has highly negative consequences on the immune system and the brain's ability to perceive available opportunities. So one's focus is to feel the positive emotions more often and using the dark side rarely, and only to lift oneself from lower energy states.

IPW Thinking Key #3: Thoughts Creating Reality

OPERATE WITH THE CONFIDENCE THAT YOUR INTENTIONS, PERCEPTIONS AND THOUGHTS DIRECTLY IMPACT REALITY

Engage life with the certainty that your every intention, perception and thought is actively shaping that which unfolds before you. This includes:

a) Assuming the emotions that you want others to mirror to you.

b) Creating intentions that you would like to see unfold in the world.

c) Focusing attention intently on the things you do want and **never** on the things you do **not** want.

Your thoughts have a tremendous impact on your movement towards or away from your IPW. This is the Pathway:

IPW Thoughts --> Behavior (Conscious & Unconscious) --> Results

When we see that pathway we are tempted - we are tempted to focus our energy on engaging in a lot of behavior/action in order to drive results. **This is a significant mistake**, and would cause us to fall out of alignment with our Ideal Parallel World. The above pathway must be approached with the understanding that IPW Thoughts lead to results. This is the key to IPW Thinking. Consider your thought as a tangible particle, that has a physical impact on the world around you. To believe in the scientific basis of this idea is not necessary, however to embrace this idea as a paradigm is critical to success. There can be no movement to your IPW without applying the paradigm that your thoughts affect your reality. To propose that your thoughts do not affect reality is to judge your consciousness irrelevant. One can debate the manner in which our thoughts affect reality but to believe that all effects begin with thought is to accept the obvious.

We shall discuss further in this pillar the reasons why your paradigm must be: IPW Thoughts --> Results

> *"If you desire a thing, picture it clearly and hold the picture steadily in mind until it becomes a definite thought-form." - Wallace D. Wattles*

All we have just outlined is the framework of IPW Thinking, now allow me to share some more specific insights on this pillar and paradigm.

IPW THOUGHTS DIRECTLY AFFECT YOUR BODY

Your DNA has the dominant impact on the structure and functioning of your body, this is well known. However it is not only the existence of certain genes that determine your physical structure and biological functioning, but additionally, it is which of those genes in your DNA are turned on and which are turned off. The process of genes being turned on or turned off is referred to as epigenetics.

> *Epi·ge·net·ic, adjective:*
>
> *". . . of, relating to, or produced by the chain of developmental processes in epigenesis that lead from genotype to phenotype after the initial action of the genes . . ."*[24] *– Merriam-Webster.com*
>
> *Medical definiton:*
>
> *". . . relating to, being, or involving a modification in gene expression that is independent of the DNA sequence of a gene . . ." Merriam-Webster.com*

Research in epigenetics is showing that a person's experiences, their external environment and the environment within their body, impacts what genes are turned on and off. Specifically, the release of certain chemicals in the brain in response to one's experiences and environment, can trigger genes to turn on or off. *This is a powerful concept as it speaks to dormant potential within all of us that can be turned on by activities within the brain. Our thoughts, can impact the expression of our genes.*

This is a remarkable new area of research that has begun to grow in the last two decades and has been more specifically titled 'Behavioral Epigenetics.'

> *"When you are inspired by some great purpose, some extraordinary cause, all your thoughts break their bonds: Your mind transcends limitations, your consciousness expands in every direction, and you find yourself in a new, great and wonderful world. Dormant forces, faculties and talents come alive, and you discover yourself to be a greater person by far than you ever imagined yourself to be."— Patanjali*

TURNING YOUR GENES ON OR OFF

The neurotransmitter (chemical in the brain) called acetylcholine, through a process called **acetylation**, can turn **on** certain genes in the body. The opposite process called **methylation**, triggered primarily by high levels of stress, can turn **off** vital genes. Behavioral epigenetics is showing that traumatic experiences, be they real or perceived, can scar and mutate the structure of DNA - this can lead to disease, and the mutations can even be passed on to our offspring. Negative experience is subjective, in that it lies in the perception of the individual – a view of the world as a place of constant struggle, and a negative approach to one's life can be the stressor that dims the light of one's genes and one's potential. Just the stress of being surrounded by failure can serve to reduce a person's genetic expression in spite of the great potential they were born with. Very much the idea that: *"For he that hath, to him shall be given: and he that hath not, from him shall be taken even that which he hath." - Mark 4:25* A poor environment can not only limit ones opportunities but can bind the expression of potential in DNA.

This is the process that unfolds, unless one ends the cycle with consciousness and IPW Thinking. Our emotional energy affects our physical anatomy. This behavioral impact on our genetic expression can be an opportunity or a threat based on how we choose to think, and then how we live our lives.

THE CERTAINTY OF SUCCESS

With admittedly circular reasoning, I say to you that by definition, it is IPW Thinking to believe in IPW Thinking. IPW Thinking is a certainty and applying IPW Thinking moves you to your Ideal Parallel World absolutely - **this must be your approach.** If you can imagine a possibility then a path exists for its achievement. Your thoughts create your reality. This is an insight that is clear to everyone who has achieved great success, and unclear to everyone who has not. But you and I are wiser than all that. Let us, just for our own amusement, joy and edification, read some quotes from great achievers on how IPW Thinking works:

> *Arnold Schwarzenegger: "The mind is really so incredible. Before I won my first Mr. Universe title, I walked around the tournament like I owned it. I had won it so many times in my mind, the title was already mine. Then when I moved on to the movies I used the same technique. I visualized daily being a successful actor and earning big money."*
>
> *Andrew Carnegie: "I am no longer cursed by poverty because I took possession of my own mind, and that mind has yielded me every material thing I want, and much more than I need. But this power of mind is a universal one, available to the humblest person as it is to the greatest."*

John Lennon: "We knew we were going to be the greatest band in the universe. We were just waiting for the rest of the world to catch up"

Jim Carrey: "I wrote myself a check for ten million dollars for acting services rendered and dated it Thanksgiving 1995. I put it in my wallet and it deteriorated. And then, just before Thanksgiving 1995, I found out I was going to make ten million dollars for Dumb & Dumber. I put that check in the casket with my father because it was our dream together".

Steve Jobs: "You can't connect the dots looking forward. You can only connect them looking backwards, so you have to trust that the dots will somehow connect in your future. You have to trust in something--your gut, destiny, life, karma, whatever--because believing that the dots will connect down the road will give you the confidence to follow your heart, even when it leads you off the well-worn path, and that will make all the difference."

KNOWING = LESS STRESS

By knowing the certainty of your future that comes with applying IPW Thinking, you have less fear of failure, less worry, and much less stress as you work from one frame of existence to the next. IPW Thinking becomes a self-fulfilling prophecy. High levels of stress become a regular state of the body if we lack a system like IPW Thinking to provide us with the certainty of achieving a desired goal. Once we have the courage to embrace the certainty of reaching our destination we are able to operate with calm, grace and poise that serves to eliminate resistance, mental blocks and high stress.

When we are stressed a hormone called Corticosterone is released which reduces the formation of new cells in the hippocampus. The hippocampus is important for numerous types of memories including spatial memory and also for categorizations of most memories. So we see that the fear and worry that occurs when we do not have a powerful belief system for achieving our dreams, like IPW Thinking - impacts memory. We learnt in Pillar I that fear and worry weaken the connection between the conscious and subconscious minds and much of this is because of the deleterious effects of Corticosterone on the hippocampus.

EFFECTS OF STRESS ON GENE EXPRESSION

We mentioned methylation - the negative epigenetic process that turns off useful DNA, triggering potentially harmful mutations that can then lead to disease. Well, no surprise, corticosterone is the primary agent of methylation. This stress effect shows one of the methods by which our fear, doubt, disbelief and the many negative emotions, can actually attract disease. The pressures of life may be real, the challenges we face may be real but the feeling of continuous stress is based on our interpretation of events and our reaction to events. The application of IPW Thinking provides the individual with a level of certainty that reduces stress; IPW Thinking provides the individual with the feeling of calm and peace and flow that reduces stress. To reduce stress is to reduce the effects of aging; to reduce stress is to reduce the impact of DNA structural damage and mutation; to reduce stress is to avoid the turning off of genes that are vital to your active functioning. IPW Thinking brings one on a path of achievement that is blissful and focused and that is perceived as effortless.

"The Master never tries to be powerful;
thus he is truly powerful.
The ordinary man keeps grasping for power;
thus he never has enough.
The Master does nothing,
yet he leaves nothing undone.
The ordinary man is busy doing things,
yet much more is left to be done." - Lao Tzu

POSITIVE EMOTIONS

The second key to IPW Thinking is to feel a dominance of positive emotions over negative emotions. With IPW Thinking we remain focused on increasing our positive emotions up the energy scale and we milk every opportunity to simply feel good. Dopamine, is the brain's primary pleasure neurotransmitter, so as we move up the energy scale, we increase the production of dopamine in the brain. The increase in dopamine levels leads to increased memory retention and learning, and this triggers the brain to release another neurotransmitter called acetylcholine.
Dopamine Levels Rise --> Memory & Learning Increases --> Acetylcholine Increases

So IPW Thinking indirectly leads to an increase in acetylcholine. The increases in acetylcholine improves mental functioning throughout the brain allowing for continued expansion of learning potential. Secondly acetylcholine leads to acetylation, which allows genes that are turned off through stress to be turned back on; this also increases the potential of the individual to adapt better to their environment by the turning on of numerous positive genes. The nucleus in each cell allows certain unused parts of DNA to be turned on by acetylation **only if** the cell needs these unused parts to better handle environmental pressure. Think of it as the same concept of your muscles becoming stronger and larger after being pushed to their limit in a workout. The muscle fibers grow back stronger (if rested and given required nutrients) after being overworked - the nucleus of each cell, like the living entity that it is, is making these determinations of what it needs to function and allows acetylation if it needs to increase performance to meet the demands of the body. So this is a real mind-body connection as the body adjusts when we choose to both challenge ourselves with the vision of an NIF and IPW and simultaneously feel positive, dopamine inducing emotions whilst doing so.

To summarize, when we feel good the brain releases more of the neurotransmitter called dopamine. As dopamine flows through the neurons in the brain it creates a chain reaction of events that causes other positive neurotransmitters, like acetylcholine to be released. This release of acetylcholine allows us to think more clearly, strengthens our conscious memory, makes us more alert and creates positive epigenetic changes throughout the body.[31]

IPW THINKING AND THE BRAIN

Our thoughts alter the brain and then our brain alters our thoughts in one continuous cycle. As we improve our thinking we directly impact the structure of our brain. This occurs as any and all thoughts cause neurons to fire along certain pathways, the more we engage specific thoughts the stronger and faster the links become along those specific pathways and the weaker other pathways become relatively. This leads to more thinking along the pathways that lead to such IPW Thoughts and the process becomes a self-reinforcing cycle.

IPW Thoughts --> Stronger Connections Along IPW Thought Pathways (Brain is changing to adjust to the increased use of circuits in those areas) ->Stronger Pathways means more IPW Thoughts --> More positive brain changes.

So for the three keys to IPW: 1. See the Future - Be the Future, 2. Positive Emotions and 3. Thoughts Creating Reality – the more we behave in accordance with each key, the more these approaches to the world become wired into our thinking. This all translates to an acceleration of our rate of progress over time. It may take us a while to learn to be in alignment with our IPW, but the more times we get our thoughts into alignment, the easier will it be to do so again and we will be able to maintain that alignment for a longer period of time. This all translates to faster progress to our Ideal Parallel World.

THOUGHTS AFFECTING REALITY - THE MECHANISM

We do not see gravity, and as clear as its effects are on our lives, there is still not a unified theory as to the exact mechanism that accounts for the gravitational force. There is this other force, that links our thoughts to all that we passionately desire and to all that we deathly fear. We must come to see the effect of this force on our lives, even though there is not a unified theory as to the exact mechanism that accounts for this force.

We must come to see the impacts of that force if we are to be aligned with our Ideal Parallel World. To use this force consciously is to engage in IPW Thinking. To use this force is to see every thought as a real entity - as a real thing. Every thought is an energy and that energy will pull you towards your IPW and into alignment, or it will push you away from your IPW into misalignment.

IPW Thinking suggests that every time we think and feel, the nature of our thoughts and feelings becomes our energy. This energy affects our behavior imperceptibly, and such leads to effects consistent with these thoughts and feelings. This energy is not a metaphysical force and it is a metaphysical force - it is in superposition. For the purpose of achieving your desired results it is much easier to view it as a metaphysical force - view it as, a sort of magic, of which only you and the ultra-successful are aware. My description of the mechanism in the subsequent paragraphs will be imprecise and potentially analogous, just as the physicist's mechanistic description of gravity is imprecise and potentially analogous. My mechanistic description of the force may be weak, but the force itself is very strong. To quote a pair of noted physicists:

> *"In our endeavor to understand reality we are somewhat like a man trying to understand the mechanism of a closed watch. He sees the face and the moving hands, even hears its ticking, but he has no way of opening the case. If he is ingenious he may form some picture of a mechanism which could be responsible for all the things he observes, but he may never be quite sure his picture is the only one which could explain his observations. He will never be able to compare his picture with the real mechanism and he cannot even imagine the possibility or the meaning of such a comparison." – The Evolution of Physics (1938); Leopold Infeld & Albert Einstein*

EXPLAINING THE MECHANISM - TRANSMISSION OF THOUGHT

In the mechanism of IPW Thoughts: All of our thoughts are transmitted. They are transmitted very literally to every cell in our body, and they are transmitted to every object that falls under our mental gaze. We are entangled with all things and our decisions as to what to think and feel have a direct impact on how we collapse the wave function of our reality.

In the mechanism of IPW Thoughts: Every thought you allow is a brain wave. A brain wave occurring at a very specific frequency. Your thoughts allow you to connect to other persons and other physical things that are on the same frequency. For example, it is very difficult to connect with someone intellectually, if the two of you are communicating on different frequencies. You may have experienced this for yourself where you say one thing, and it is interpreted by another as something entirely different. Where maybe you have tried to send out a positive word, and someone on a different negative frequency has interpreted your well-meaning thoughts and words as something malevolent. Of course, the inverse is true, there have been times when others have tried to reach out to you, and you have misjudged or misheard, and misinterpreted their intentions. It was because you were on different frequencies with them. Your thoughts were aligned to different things, and your very view of the world, was different than this other individual's view of the world, at that moment.

Along these same lines, it is very difficult to accomplish a goal or to achieve anything of significance if you do not align your own thinking to the frequency of that goal or accomplishment. You must be on the same frequency as your Next Ideal Frame and Ideal Parallel World, this is 'See the Future – Be the Future.' To be on the same frequency as your IPW you think and act 'as if.' You think 'as if' you had already achieved the great success – you think 'as if' you were already the great actor, 'as if' you were already the great lawyer, 'as if' you were already the great CEO. From this moment forward, you decide to think

As if You were already ______________________________

As if You were already ______________________________

As if You were already ______________________________

As if You were already ______________________________

To be on the same frequency as your IPW, do not wait for the day when you will have the ideal virtues; instead assume that you have these virtues now, and think from that perspective. Use your 'I Am' and allow yourself to Be the Future:

I Am confident.

I Am positive.

I Am loving.

I Am grateful.

I Am joyful.

I Am __________________________________

I Am __________________________________

I Am __________________________________

I Am __________________________________

In the mechanism of IPW Thoughts: Each of these 'I Am' thoughts and 'As if' thoughts are real particle expressions and by choosing appropriate thoughts, you can make these parallel shifts in reality that will allow you to Achieve Your Dreams. These IPW Thoughts are how we shift the flow of current reality and cause the wave function to collapse in any manner we desire.

> *"And God said unto Moses, I Am That I Am . . ."*
>
> *- Exodus 3:14*

THE PROOF

There can be no proof for IPW Thinking other than your own experimentation and future experiences. And that is what your life is all about. This is what your personal evolution is all about. It is an early lesson in learning to use the power of the force that lies within your own mind. This experience on Earth is, in a manner of speaking, an introductory course on learning to use your words to create your own Ideal Worlds.

> *"Through faith we understand that the worlds were framed by the word of God, so that things which are seen were not made of things which do appear." – Hebrew 11:3, King James Version*

If what you are doing in your life is serving you well and continues to work then you should persist along that path. If what you seek is a greater power, a greater influence and a real ability to wield an infinite force for creation – then the path is through your thoughts.

YOUR THOUGHTS

Every path to your Ideal Parallel World is based on what you allow yourself to think. The programming of the subconscious mind we discussed is about bypassing years of thought and engaging in the inception of new thoughts and this pillar, IPW Thinking, is all about thoughts. Everything is Thought. Where you are today is because of all the things that you have thought – you may accept this now, or you may wait until you are hugely successful to accept this – but the longer you wait to hold your thoughts accountable, the longer it will take you to arrive at your Ideal Parallel World.

Every moment we exist, every moment we think, we are broadcasting all our 'thought waves.' These thought waves are what created our present and will be what creates our future. Our thoughts alter the behavior of those around us, because our thoughts alter our own subconscious behavior. Our thoughts are drawing to us objects, activities, people and happenings, as well as other thoughts, on similar frequencies. If we place thoughts of love in our mind we will draw love to us, if we think on hate we shall create more hate, if we think gratitude we will draw gratitude, all manner of thoughts be they empowering or be they of the dark side, will be drawn to us on par with what we have broadcast to the world through our thoughts. This process is not magical, necessarily, and with Earth as your introductory course, this process is not often instantaneous. It is just as unreasonable to have your first course in thought power be instantaneous, as it would be unreasonable to have a baby's first course on walking be on a 10 foot high tightrope.

"With great power comes great responsibility," and as we evolve and become more aligned with our IPW and, in a manner of speaking, with our higher selves, our power grows, arithmetically and then exponentially. IPW Thinking shall not be instantaneous just yet - you may sow a thought of hate and it may take years before you experience the effects of such hatred. You may sow a thought of commitment to quality at work (in spite of an ungrateful boss), and it may be months or even years before you reap the harvest. The harvest is not always exactly what you expect, but it will be exactly what you need to advance to your Ideal Parallel World. Know that the thought is a real thing and therefore you must choose your thoughts wisely.

WHERE TO FOCUS?

Having our minds pursue all manner of interests can be quite dangerous. A compelling focus on your Next Ideal Frame and Ideal Parallel World are necessary to achieve success. Until your understanding of the force is infinite then your energy will not be infinite. So working with the store of energy you have as you evolve, you must expend energy wisely, through focus.

To see love in the world is to be on that same frequency as love. To see and acknowledge a thing is to be on the same frequency as that thing. That is why we are called to focus on that which is good in even the worst of circumstances. By so doing we grow that which is good; we do not curse the darkness but we shine a light – we expand the reach of the light, not fight the darkness. This must be our approach to all things:

As a manager we would focus our efforts on celebrating great employees and rewarding great behaviors and spend an almost negligible amount of time condemning, criticizing, punishing and complaining. We can apply the rules and required reprimands to those who do not follow the standards, but we do not dwell on the darkness. In a sense, though not literally, it is not that 'we do not suffer fools gladly - ' but more accurately, it is that, we do not suffer fools at all.

As a teacher, we would focus on rewarding the good student and not focusing such tremendous effort on maligning the bad student.

As a parent, we would dwell on and celebrate the positive behaviors of our kids. Whilst simultaneously ensuring that their unpleasant behaviors are not rewarded and do not garner the type of attention that the misbehaving child may be seeking. We replace attending to the negative, with a focus on the positive.

As a student, we would focus on achieving breathtaking performance in the subjects where we are strongest whilst avoiding the inevitable mediocrity that results from focusing on areas where we are weakest. Improve your weaknesses, yes, but focus powerfully on your strengths.

As a lover, we focus our eyes on what is beautiful in our partner and not on what is less beautiful. When we are in love we do this automatically - our eyes dilate to take in every visible frequency of light projected from their visage. We see the other person in light of their best intentions and by doing this we recognize our own beauty and our own light. Henceforth, this shall be your manner with all things.

Keeping our thoughts on the positive frequencies in all circumstances is a concerted effort to grow that which is good instead of using our best energies to fight that which is less good. Mother Teresa understood this well when she said:

"I will never attend an anti-war rally; if you have a peace rally, invite me."

THOSE WHO DOUBT

You shall notice, as you become more and more aligned with your IPW, that the haters in your life become irrelevant; the words of the doubters and dissenters carry no weight; the attempts to malign you cause you to chuckle, and a word from another, suggesting that you are weak, causes you to laugh uproariously.

To respond to threats, and obstacles and daggers with the thought: "Surely, you must be joking. Do you not know who I Am?"

To see the wall in the distance and say, "It must not know that I Am on my way – for there is no force that can stop that which I Am."

LIKE ATTRACTS LIKE

> *"You are a human magnet and you are constantly attracting to you people whose characters harmonize with your own."* - *Napoleon Hill*

A consequence of non-IPW Thinking, that is, negative emotions and negative thoughts, is that a release of negative thought waves creates a comfort and familiarity with such waves for the individual. This causes one to then see more of that which is on a similar frequency. In psychology this is called 'mood-congruent thinking.' Our memories each have an emotional component to them and whenever we feel an emotion the brain conjures up similar memories and experiences that align with that emotion – then we begin to see our entire world all through the lens of that emotion, be it for good or ill. So because of this matching of frequencies and mood congruent thinking, not only would one be attracting negative frequencies from everywhere, but one also becomes quite adept at recognizing these negative frequencies, and so it starts to seem like the entire world is in fact negative and in unstoppable decline -- this creates a vicious cycle of more negativity.

Beautifully and fortunately, the inverse works as well: by choosing to feel positive emotions and focusing on positive thoughts, we create a comfort and familiarity with such positive thought waves. We then attract to us similarly positive thought waves and we become more adept at identifying the positive -- again it is a vicious and pleasant cycle that we must foster to keep us in alignment with our IPW. The path to alignment is the path of positive emotion. It is the path of bliss, love, joy, peace, allowance, gratitude, empowerment, freedom, confidence, passion, enthusiasm, belief, engagement, eagerness, optimism, hope, humility, compassion and more. Just saying these words lifts the spirit, and implants briefly onto the subconscious mind.

IPW THOUGHTS AND YOUR ENVIRONMENT

You now know, that every element of your present life is a manifestation of your past thoughts. To see the nature of your past thoughts all you need to do is look at where you are now and look at how you exist in this moment. There are many things in your life that are going very well, and there are many things that you still desire to accomplish. To move forward, you must see that this moment, the nature of your existence, is a three-dimensional holographic representation of all your past thoughts and feelings.

Admittedly, that which you are, this present manifestation of your past thoughts, has not occurred in a vacuum. You do exist surrounded by the energies of others, friends, family, work colleagues or academic peers. However, the manner in which this environment, affects you, has more to do with the cyclical effects of positive versus negative emotions that we recently mentioned. That is, if you are in a highly negative environment and choose to embrace negative thoughts, the rate at which you will receive vibrations on that negative frequency will be faster than if you were in a positive environment.

Imagine a kid growing up in a crime ridden neighborhood with thieving neighbors, and drug dealers and prostitutes on every corner. If that kid wants to be a great physicist and a model citizen, then she really has to stay on a very narrow path of studying daily and focusing at school, and eating healthy and being home before dark, and not stepping off the path of discipline and good behavior. Because if that kid decides to experiment with drugs, or sex, or stops studying, there will be a chorus of her friends and neighbors who would reinforce that behavior down that dark path - sort of a regression to the mean (average) level of negativity. So, in the negative environment, theoretically, she can still accomplish any great thing she wants to, but only if she remains hyper focused - for if she strays, but a little, she can fall very quickly and may never recover.

Now put our same hypothetical kid in a neighborhood where her neighbors are successful, the couple on the corner are scientists, the drug use is all hidden and prescribed, her school is well equipped, her teachers are inspiring and effective, a successful community, with many successful people - a positive environment. If she strays off the path to becoming a great physicist and model citizen, firstly she will feel a greater push from her friends and teachers and neighbors to get back on the right path, but secondly, even if she suddenly became a mythical demon child, because her environment is reinforcing the positive and not the negative, then her decline would be much slower than in the negative neighborhood scenario. The average level of success is higher in this second scenario, so her regression to that mean is not at all a tragedy.

Similarly, if you are surrounded by an abundance of negative energy and begin with your own thoughts to move over to that dark side, your rate of acceleration along the darkened path will be faster than if you were in a positive environment. We see that the external environment including your family, your friends, peers, work colleagues, etc., impacts your own reality, but still the ultimate responsibility does lie with you, in your choice to also broadcast your thoughts on negative frequencies, and your choice to accept others' thoughts on negative frequencies. To be positive, and to have IPW Thoughts, in such an environment is difficult, but clearly it is the work one must do to alter their path.

If you believe in what you are doing, are tapping into your subconscious mind as you should be, have created an image of the IPW and are consistently visualizing and then feeling the emotions of having already arrived at your IPW, then the thoughts of weakness, or failure, or any negativity of those around you should have a less dramatic effect. The world behind your eyes has become more real than the world in front of it.

No matter our environment, when we dwell on success, when we dwell on that which is good, on that which is positive, we attract success, we attract that which is good and we attract that which is positive. We connect well with others when we get on the same frequency as them.

My friend David spent almost two years working for a company where in spite of his ability, he struggled to do well and never got the promotion he was working towards, but most importantly he shared with me shortly before he left, that he often had the feeling that he did not fit in. I told him, to his surprise, that he was exactly right. Out of more than eighty employees, there was probably myself and two others, who were on the same frequency as him. He left that company, and left that environment, and moved on to other areas where he proved to be much more successful. And this is one possibility if we happen to find ourselves surrounded by frequencies and broadcasts that do not align with our own. By maintaining our focus on the frequency of our IPW, we come to quickly realize whether a current environment will serve us or hinder us and we may choose voluntarily or involuntarily to be expelled from that environment. See that expulsion not as a negative, see it as an ascension from a lower plane of consciousness to a higher plane of consciousness. See that expulsion as movement from a less positive section of the thought matrix to a more positive section of the thought matrix.

YOUR FAMILY, FRIENDS AND PEERS

Accept that the people you spend the most time with will have a tremendous impact on your life. The mechanism by which this works is through the reinforcement cycle. If the energy of your peers is not on the frequency aligned with that of your IPW then their thoughts (manifested through their actions and reactions) will create a mass of potential energy that could push you away from the path to the Ideal Parallel World (IPW). The mass of negative potential energy becomes kinetic energy, and initiated, only if you begin to doubt yourself. If you are carrying out the daily exercise of planting the right thoughts and ideas into your subconscious, then your own level of certainty and alignment to the IPW will be high and this will limit the negative impact of others. Because your environment can feed the negative or positive cyclical effects it is a safer strategy to place yourself in environments that will serve to reinforce you positively.

THE FIVE IN YOUR ENVIRONMENT

A key step in shaping your environment, is to look at the five people with whom you spend most of your time. In life you will find that the five people you spend the most time with will have the greatest impact on your life. Over time you will become an average of them in traits and career and life success within one or two standard deviations. This is a simplification that says what we have already stated, that is, your environment powerfully influences your reality. This concept of, 'you being the average of the five' is a very useful measure of the nature of your environment.

Create a list of the five people you spend the most time with every week. This may include family, friends, schoolmates, work colleagues, anyone. Your interaction with these people may well not be a choice you made consciously, for example at work, but their influence on your life occurs nonetheless and must therefore be considered.

<u>My Current Five</u>

Person One ____________________________________

Person Two ____________________________________

Person Three __________________________________

Person Four ___________________________________

Person Five ___________________________________

Understand that you are now, or will become, the average of those five people in one form or another – not just in material success but even more subtle attributes like drive and self-confidence and humility. A friend said to me that he spends most of his time with his infant son and his wife so by now he should be drooling, babbling and doing yoga, but it is at a much deeper level than this. Being with an infant for extended time stretches can teach us very profound lessons about growth and passion and curiosity, creativity and trust. All of which are valuable traits that we may want blended into our own. The act of parenting makes many men more patient, considerate, nurturing, resilient, generous and numerous other positive traits linked to the time spent with both mother and child. So this simplified 'rule of the five' does continue to hold.

The question you may ask yourself is: "Do these people inspire me, support me, challenge me, push me to do more and have more and be more?" Go ahead and make an executive decision on who you should maximize your time with. Do not focus on eliminating anyone or noting the negativity of a certain person, because by so doing you put yourself on a negative frequency. Instead what you must do is focus on the persons you **do** want to spend more time with and find reasonable ways to do this. By focusing on the positive and by focusing on increasing your interactions with those key positive people you will automatically crowd out all else. This will be your source of light filling the room and eliminating the darkness. Remember, this method of focusing on the positive over the negative is a key part of IPW Thinking. You may edit your five, if needed.

"The key is to keep company only with people who uplift you, whose presence calls forth your best." — Epictetus

<u>My New Five</u>

Person One ________________________________

Person Two ________________________________

Person Three ______________________________

Person Four _______________________________

Person Five ________________________________

Those who you are closest to will alter your thinking, your view of the world and your focus and powers of awareness. There is no way around this and this is not a judgment, nor is this saying to trade your friends and significant others as though you were the manager of a sports team - or maybe it does suggest this? It is your call to make, but it is difficult to win if you are surrounded by ~~losers~~ negative frequencies. Be aware that your closest people are having this effect. Your closest five should believe in your potential and should grow to support you and you should also grow to support them.

Interacting with others who do not see you as you see yourself serves to *collapse your wave function* in a manner that does not serve your advancement to your IPW. We have all had great ideas that we got really excited about, and then tell a friend who then says 'it can't be done,' and the blood drains from our cheeks and our energy is sapped – that is allowing someone else to *collapse your wave function*. That person outside of yourself has removed you from the sea of infinite potentialities that we all exist in, that person has removed you from the wave flow of possibility and caused your reality to become a single particle expression of weakness. Never let this happen. People who do not believe in your potential can pull you down to a reality that is below your potential. They may well be trying to make you come back to reality – to their reality! Do not do it. F*** their reality. Achieve greatly by seeing your own reality and being your own reality.

Reality, is in a sense, a mutual agreement of minds. Your mind needs to be strong enough so that your view of the world impacts their reality instead of their view of the world impacting your reality. This is not about being confrontational, this is about knowing who You Are, knowing your IPW and existing in the certainty, peace and serenity of inevitable success.

When someone is clear about their 'I Am' and walks into a room, everyone in the room, consciously or unconsciously gets the feeling of an elevated energy. One person walking into a room can change the energy of the entire room for good or ill.

Your responsibility is to be that individual who is operating on a higher frequency, and raising the energy of all who you come in contact with.

We can think of an example as when a person is courageous in a dangerous situation whilst others in the group are fearful, that courageous person will either increase the feeling of courage by the other group members or he will succumb to fear himself. It is rare that one person's energy remains on a high frequency whilst everyone else's remains low. There is most often a mingling of those energies to an approximate average of the group.

> *"If you can keep your head when all about you, are losing theirs and blaming it on you . . ." - Rudyard Kipling*
>
> *"One man with courage is a majority." – Thomas Jefferson*

This is where you must be with the energy you feel of already existing in your Ideal Parallel World – you must feel the higher energy, that is, the higher frequency thoughts and emotions of your Ideal, and you must maintain this higher frequency in spite of the frequencies of those around you. By being consistently on a higher frequency others will gradually begin to turn to you for wisdom, advice and leadership. They will follow your lead and begin to behave more in line with your higher frequency and will begin to see you more as you see yourself. This impact of your energy on the energies of others will serve your movement to your Ideal Parallel World.

PATIENCE IN A NEGATIVE ENVIRONMENT

Never have fear or worry about your current environment. As fear and worry only serve to begin the cycle of negative attraction which will be rapidly amplified by this same environment. You may find yourself in a work environment filled with negatives. Your responsibility is to focus on the positives, the strengths, that which can be learned from and used to grow within that environment and eventually outside of it.

IPW Thinking works very closely with your expectations, and you will attract to you that which you feel most strongly about. If the strongest emotion you feel during the course of your day is one of fear and worry, then this will serve to attract more of the types of circumstances that led to your fear and worry.

To ruminate over the negative aspects of your current environment is to attract more of a similar environment. If you fear being stuck in this negative environment then you will be stuck in that negative environment - quite simple. If you focus on that which you can learn and the positives within that environment, as well as creating an image of the type of environment you intend to be in, then you will learn from your current environment and you will take the lessons from your current environment into the ideal environment that you have imagined. That new ideal environment can be your Next Ideal Frame or maybe that is even your IPW, but I suspect your IPW is much grander.

To avoid fear and worry you must plant ideas of strength and courage in your subconscious mind through the methods we outlined, and as part of your daily ongoing thoughts. You must assume the feeling of courage and strength. As with all things, cursing the darkness is not a path to your NIF & IPW, only expressions of light, that is, only the positive spectrum of emotions involved in IPW Thinking is an acceptable path.

Your thoughts must be of courage and capability:

"We must build dikes of courage to hold back the flood of fear." - Martin Luther King Jr.

So long as you have fear, you are stuck, you are done, all paths to your Ideal Parallel World are improbable - let that be most clear. To understand this, is to understand why "*the only thing to fear is fear itself,*" but frankly, do not even fear fear - have courage so great that when you enter a room fear leaves. Implant ideas in your mind that your courage and strength are so great that fear is terrified of you.

Can you imagine how the fear of failure cripples so many great enterprises? Can you imagine how many adventures and ideal worlds go un-pursued due to fear? How crippling it is to sit back and contemplate the possibilities of failure and say "maybe it won't work, what if this happens, what if I look foolish, or maybe I should or maybe I could, maybe this, maybe that?" Fear is a breaker of backbone. Fear is the destroyer of spine. Fear is that which disconnects your actions from the will of your higher self, your IPW-self. Fear separates you from infinite possibility.

As an ongoing exercise, make a habit of doing something which you fear at least once a week and do something that makes you uncomfortable every day. The result of all this - the result of building courage, is to strengthen the connection between your actions and your higher self. The result of this is to strengthen the connection between your actions and infinite possibility. Once you implant courage into the subconscious mind and courage becomes your way of life the link between your current frame and your IPW is strengthened and your vision of the path from your current frame to your IPW becomes clear. Fear clouds vision.

The greatest battle you must undertake, the foremost battle you must wage, the most important battle that there is, is the battle of courage versus fear – and there can be only one victor – that victor must be you and courage, or it shall be the dark side and death.

FEAR BY ANOTHER NAME

Be sure to recognize fear in all its forms. Fear is often disguised as worry. As such, worry is equally dangerous. I know you may think on times when fear and worry drove you to complete a task, maybe drove you to meet the deadline for an important project, or maybe made you work even harder to achieve certain financial goals - you are correct in that perception. The challenge here is not that fear and worry are entirely unproductive, the challenge is that fear and worry are of the dark side and are a path to the dark side. They lead you to a place where you desire not to be. They bring with them a whole slew of negative emotions that are characteristic of the dark side. The dark side is the path of greatest stress, and a path that creates enemies and opposition. Indeed, the dark side can deliver quick results in the very short term, but it hardens every cell, every ambition, every fiber of your being. The dark side stiffens the body and stifles future achievement and possibility. There is no path to the Ideal Parallel World through fear and worry. The dark side cannot take you to the IPW and even if it could take you there, such achievement would be unsustainable. The dark side, the dark emotions consume all and destroy all. You may be able to wield the dark forces for a moment, but only for a moment and to your own peril, and to the peril of those closest to you. To compel yourself to action do not use fear or worry, but instead use your confidence, use focus, use gratitude, use love, use the positive emotions. Enveloping those emotions around your Next Ideal Frame will create the energy to drive accomplishment, and so wielding the dark force of fear becomes unnecessary.

> *"You can never become a great man or woman until you have overcome anxiety, worry, and fear. It is impossible*

for an anxious person, a worried one, or a fearful one to perceive truth; all things are distorted and thrown out of their proper relations by such mental states, and those who are in them cannot read the thoughts of God."- Wallace D. Wattles

THE LIMITS OF YOUR POWER

In the early stages of understanding and wielding the power that is available to you, you are working within certain energy limitations. Let me be clear, the power of the subconscious mind and the infinite potentialities of the wave function connect you to an infinite power - but until your understanding of these forces becomes infinite, then your energy and power shall not be infinite. So as a physical being you have a reserve of energy that you have access to every day, so you must decide where to focus this energy; you can focus it on that which you can grow and on that which you intend to do well, or you can quite erroneously focus it on worry and fear. Worry and fear produce nothing, whereas a focus on ones Ideal Parallel World with clear positive intentions, produces all things. So invest wisely - this is not optional - there is only one path.

ALLOWANCE

As part of IPW Thinking you must be willing to embrace the world where you do not achieve you desired ambition, yes; imagine confidently being able to handle the consequences of non-achievement. This must be done with an emotion of calm courage. Paradoxically, this feeling of surrender will give you the strength to achieve your NIFs and will serve to reduce fear.

To be willing to surrender is to be fearless. Willingness to surrender is actually very courageous, because it is an awareness of one's ability to psychologically cope with all possibilities. In order to win, one must be willing to lose.

> *"Let your soul stand cool and composed before a million universes." – Walt Whitman*

THE ILLUSORY ARROW OF TIME

> *"The separation between past, present, and future is only an illusion, albeit a very convincing one." – Albert Einstein*

Einstein's theory of relativity suggests that time is an illusion. The straight line we perceive of one thing happening after another is not real. It suggests that all things are happening in reverse order, in standard order as well as simultaneously. This is not at all what we see as we go about our lives, so the brain must be creating for us, our perception of linear time. Again we start to see that the behavior of the brain is much more like that of a computer, programmed to create this illusory perception. But even if we accept this warped view of time, how would this really help us achieve any goal? How does this knowledge allow us to move to our Ideal Parallel World? Here's how this relates to IPW Thinking:

Embrace for a moment the idea that the timeline we experience is an illusion. To embrace this idea, one would have to also reject the idea of cause leading to effect. If time can move in both directions then effects can precede causes as frequently as causes can precede effects. So in a linear/straight line view of time you may achieve a goal and then feel good. In a non-linear view of time you will need to feel good about achieving a goal and then you will achieve the goal.

Old Paradigm: Achieve Goal (cause) --> Feeling Happy (effect)

***New** Paradigm: Feeling Happy (effect) --> Achieve Goal (cause)*

This is very much what you are doing when you: 'See The Future – be the Future.' You are taking on the mental state of having already arrived at your Ideal Parallel World in order to arrive at your Ideal Parallel World.

The consequence of non-linear time is that: Achieving your goal must not be what causes you to feel the stream of positive and empowering emotions, but instead it is the effect of feeling the stream of positive and empowering emotions that will cause you to achieve your goal. The effect precedes the cause. *"Success does not lead to happiness, but instead happiness leads to success."* Yes, success can lead to happiness - the point is not that time goes backwards or that effects always precede causes – the point is that time exists in both directions. So the new evolved method of human perception of time must be that the order of cause and effect is multi-directional.

Old Paradigm: Make $1Million (cause) --> Feeling Happy (effect)

***New** Paradigm: Feeling Happy (effect) --> Make $1Million (cause)*

This is not a complicated change and I suggest that you have already been applying this multi-directional view of time in your own life, but may not have perceived it as such. If you have ever treated someone with love with the goal of having them love you back, then you have applied 'effect --> cause' approach. In that example you did not wait for them to love you and then love them - you started out where you wanted to end up. There may be other times where you were timid in romance and were hesitant, and were reserved with your expressions of love, and the result you got reflected back was reservation and timidity, which led to a cycle of gradual reductions in the closeness of the relationship or a prolonged friendly progression of non-intimacy - friendzoned. Seeing effect before cause is what we do when we pursue a goal by beginning with the end in mind. In the areas of our life where we are most successful we do this best. We visualize the end goal and we know exactly what we intend, and then we consciously and subconsciously create with an ease and certainty that we may take for granted. This is what the process of effect before cause is all about, and so even though at first glance embracing time as an illusion seems extreme, in some ways, you are already doing this. You must continue to do so in all areas. To receive peace act peaceful; to win friends be friendly; to win support be supportive; to be understood focus on understanding; to become the boss think like the boss; to become the leader, dress, think, act and strategize like the leader.

The important step here is to acknowledge the paradigm of time as an illusion, and to realize that to set your emotional state now as though you had already achieved victory is part of the path to victory. If you dwell on the common phrases 'thinking like a winner' or 'playing to win,' one sees further anecdotal references to this paradigm in action. 'See the Future – Be the Future!'

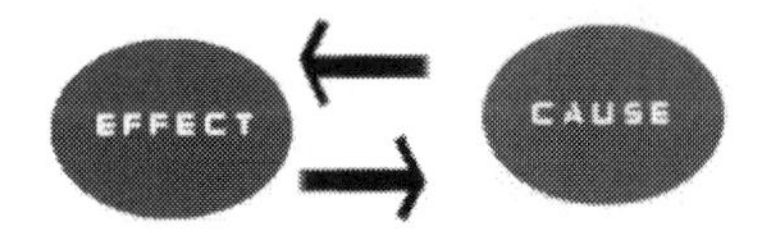

THE ILLUSION OF TIME, AND CREATING REALITY

We are creating this 'reality' of the physical world through our emotions, our energy and our thoughts. The reason we are normally unable to decipher this process is because we previously believed that cause lead to effect, instead of effects leading to cause. So by focusing on the effect, which is our feeling, we can create the cause, which will be our Ideal Parallel World.

THE DREAM

Tonight, you fall into REM sleep and a dream begins. You are dancing at a smoky nightclub and eating a blue waffle and simultaneously trying to finish a project for your angry boss who is in the form of a Lion. You finish the waffle, sneak out the club, escape the Lion attack, and you find a quiet spot, and pick up a book. You start to read. The first page of the book says in bold letters, "YOU ARE DREAMING," you turn the page again, and it says, "YOU ARE DREAMING." Every page in the book just says that "YOU ARE DREAMING." Do you believe the book?

If you do **not** believe the book, then you are resigned to continue your life of occasional hurried dancing, and blue waffles that you do not get to fully savor, and constantly looking over your shoulder worried about the teeth of Lions. It is a decent life - we are all very fortunate, as there are worse things.

If you **do** believe the book then you have the opportunity to really become a key player in the dream world. Being aware of the illusion at work, you can progressively learn to harness the power of your thoughts and start the shifts in reality that give you an eventual abundant ability to manifest. A power to Think 'I Am ....' and silently watch the world contort itself to create a coherent structure that matches that which you insist shall be your destiny.

"A wise man, recognizing that the world is but an illusion, does not act as if it is real." - The Buddha "A wise man, recognizing that the world is but an illusion, does not act as if it is real." - The Buddha "A wise man, recognizing that the world is but an illusion, does not act as if it is real." - The Buddha "A wise man, recognizing that the world is but an illusion, does not act as if it is real." - The Buddha "A wise man, recognizing that the world is but an illusion, does not act as if it is real." - The Buddha "A wise man, recognizing that the world is but an illusion, does not act as if it is real." - The Buddha "A wise man, recognizing that the world is but an illusion, does not act as if it is real." - The Buddha "A wise man, recognizing that the world is but an illusion, does not act as if it is real." - The Buddha "A wise man, recognizing that the world is but an illusion, does not act as if it is real." - The Buddha "A wise man, recognizing that the world is but an illusion, does not act as if it is real." - The Buddha "A wise man, recognizing that the world is but an illusion, does not act as if it is real." - The Buddha "A wise man, recognizing that the world is but an illusion, does not act as if it is real." - The Buddha "A wise man, recognizing that the world is but an illusion, does not act as if it is real." - The Buddha "A wise man, recognizing that the world is but an illusion, does not act as if it is real." - The Buddha "A wise man, recognizing that the world is but an illusion, does not act as if it is real." - The Buddha

Very few of us engaged in the dream would believe that book - nor must we, as merely considering that reality is an illusion, is enough of a seed. Just reading the pages of that book would count as the inception of an idea. Reading that book, as a co-creator, would count as hacking into your own mind, and planting the seed of the idea that this world is holographic and illusory. Just reading the pages of that book, would make you more aware and more powerful, now, than at any other point in your existence. You are more powerful, now.

Even still, the problem with cracking the illusion of a dream is that the very brain that is creating the dream is also creating the logic and evidence to override the realization that it is a dream. Lucid dreaming and lucid living rarely occur on the first attempt. Just as learning lucid dreaming may take many nights, achieving enlightenment, that is, mastering lucid living, may take many lifetimes.

We established that one of several reasons why you may find it difficult to decipher that this is a dream, is because you have believed that cause lead to effect and never the other way around. But remember, the new idea which is a key part of IPW Thinking, is that effect can lead to cause:

Emotions/Thoughts --> Physical Reality

Consider: In a dream, your subconscious thoughts are what create the dream world because the conscious mind is basically sedated. Then, is it really odd, that in this conscious reality, the opposite effect is at play? That is, your conscious thoughts and feelings are the creator of your physical world (the conscious communicating to the subconscious), just at a more delayed rate.

Effect (Feeling: 'As If,' Happy, Confident) **--> Cause** (Achieving Goal)

I appreciate that I may be pulling you far from the shore right now, but stay with me as we go further. As we consider the Holographic Universe, we also consider that each of our realities is entirely subjective. We exist in our own minds, seeing reality from our own unique perspectives, essentially in a world of only ourselves. In a sense, every other person and organism, every plant, is in fact your own consciousness manifested. The best analogy to explain this is that of a typical nightly dream – in a nightly dream, there are other characters who you interact with but essentially they are all you, because the entire experience is occurring in your own mind. The characters in your dreams are represented as people separate from you but they are a manifestation of your own consciousness. The physical world is like this, except that the structure of the physical world is denser than that of the dream world. The ability to manipulate the physical world is real, but the manner of manipulation is different than the dream partly because the result of manipulations are delayed and seemingly reversed (effect before cause; conscious before subconscious). That is, the real world responds directly to your thoughts but at a delayed rate compared to the manipulations that one may engage in lucid dreaming.

Now everybody calm down, breathe deeply and put your clothes back on. The suggestion that this is a dream is an idea in superposition. However the paradigm that your thoughts, emotions and consciousness are creating your reality, is essential for guiding your movement to your Ideal Parallel World.

Timeline for Success

The quantum particle in superposition that allows you to split reality is in your own consciousness. When you become aware of the frame you are in, and take a breath and take a snapshot of your existence, at that point there are an infinity of decisions you may make on how to proceed. That moment of awareness, that moment of conscious choice, that moment of an IPW Thought, is the point where the universe splits and you move towards your destiny.

Michael Jordan once stated that when he is at his best, the defense seems to slow down and he is able to create opportunities for scoring with greater ease. This is the shift in perception that occurs as we become more conscious and narrow our focus onto our Ideal Parallel World. In each moment of consciousness the world slows and we can see more frames per second, per minute and per hour. By definition, to focus, is to attain more detail along a narrower band - your Ideal Parallel World becomes that narrow band and focus gains you access to more frames of awareness in each moment.

Here is the clearest answer you will receive on manifesting and achieving goals within a time frame. A clear answer to the question: "How long will it take to get to your IPW?" As you will observe this answer may be clearer than most but may be of little comfort to many. Here we go: Any IPW that you imagine can be achieved the instant you imagine it, however the greater the ambition relative to the current positioning of the current elements in the frame, the lower the probability of instantaneous achievement. Stated differently, it is within the probability wave of the frame of life you find yourself in, that you can instantaneously achieve your goal; but instantaneous achievement is just less probable than achieving a goal through progressively closer approximations, one frame at a time, towards your Ideal Parallel World. All this means is that time must be your friend if you want to achieve a shift in your frame significantly discontinuous from your current path. A particle can be where you expect a particle to be or it could in fact be anywhere else in the physical universe, but that distance is impacted by probabilities. The particle is more likely to be 1 meter away from the expected position, than is it to be 100 meters away. Each time you become aware, you can move the elements of your frame so that you are closer to your IPW and by so doing there is a greater probability in the next frame of instantaneous achievement of your goal. Every time you move closer to your IPW, the probability of instantly getting there increases.

One must make alterations to physical frames to increase the expediency of success. To accelerate your progress through frames is to become stronger at any of the competencies described in the 2 Pillars. For example, by becoming better at influencing your subconscious mind through 'auto-suggestion' and self-hypnosis, you can accelerate your progress through the frames; by more rapidly increasing your ability to show gratitude and experience love as part of IPW Thinking you will have more energy to be aware and to manipulate a greater number of frames in a shorter period of time. The more frames you actively manipulate and the more frames where your thoughts are IPW aligned, the fewer frames it shall take you to arrive at your Ideal Parallel World. To become aware of your entire self, beyond just the ego (which is the conscious mind) is to achieve an awakening – a heightened sense of awareness that allows the individual to see more frames per second/minute/hour and therefore move through frames at an accelerated rate relative to one's past self and to others. To feel a sense of purpose and mission in life is to feel a sense of awareness, and connects you with your higher self and access to the subconscious. Self-awareness creates speed and inevitable success.

The subconscious mind chooses how to collapse the wave function to provide you with a single objective reality. It does not usually manifest your Ideal Parallel World instantaneously because it also must work through the probabilities and possibilities that are inherent in your current frame. If you are standing on Mars without a space suit and oxygen tank the probability set available to your subconscious mind for ways of keeping your body alive will be very limited and you will almost certainly die - the subconscious mind is not a miracle worker, it is more like an immeasurably powerful processor. A processor more than 2000 times better at handling information and **more than** 20 times better at strategizing and calculating variables than is your conscious mind. But as you can imagine, using a processor that is so many times more powerful than your conscious mind will feel like working miracles. You will begin to feel all powerful and the process of creating a physical world to suit your fancy will become simple. Remember:

> *"Where you are today is because of all the things that you have thought - the longer you wait to hold*

> *your thoughts accountable, the longer it will take you to arrive at your IPW."*

This is not magic, but it is magical. Everything you have read in this book is meant to create a paradigm shift in your own mind. It is meant to allow you to see the world differently and see your current reality for the illusion that it is. Indeed, you are working through a wave of probabilities, but the application of the 2 Pillars allows you to collapse that wave of probability in a way that seizes the ideal possibility.

Thank You for helping me create this, my first book. ☺
Feel free to email me questions/comments/feedback:
KevinLMichel@Gmail.com.
If you enjoyed the book I would greatly appreciate a review on Amazon. Best Wishes!

Moving Through Parallel Worlds To Achieve Your Dreams

Sources

1. Og Mandino (1968). The Greatest Salesman in The World.

2. Napoleon Hill (1928). The Law of Success, Lesson 4, The Habit of Saving.

3. Hugh Everett III, (1956). Thesis: Theory of the Universal Wavefunction. Princeton University.

4. The Matrix Reloaded (2003). The Wachowski Brothers (Screenplay).

5. Neville Goddard, N. (1944). Feeling Is The Secret.

6. Rhonda Byrne (2010). The Power.

7. Alex Lickerman, M. D. The Undefeated Mind: On the Science of Constructing an Indestructible Self.

8. Neville Goddard (1941). Your Faith is Your Fortune.

9. Norman Doidge, M.D. (2007) The Brain That Changes Itself: Stories of Personal Triumph from the Frontiers of Brain Science.

10. Napoleon Hill (1928). The Law of Success in Sixteen Lessons.

11. Deepak Chopra (2010). The Soul of Leadership.

12. Caroline Myss PH.D (1996). Anatomy of The Spirit.

13. Rhonda Byrne (2006). The Secret.

14. Christopher Nolan (2010). Inception (Screenplay).

15. Gould E., Tanapat P., McEwen B.S., Flugge G., Fuchs E. (1998). "Proliferation of Granule Cell Precursors in the Dentate Gyrus of Adult Monkeys is Diminished by Stress".

16. Abraham Hicks(2013). Live Seminars. Law of Attraction.

17. Champagne, FA; Curley, JP (2005). "How Social Experiences Influence the Brain". Current opinion in neurobiology 15 (6): 704–9. doi:10.1016/j.conb.2005.10.001. PMID 16260130.

18. Gary Zukav (1989). The Seat of The Soul.

19. The Many-Worlds Interpretation of Quantum Mechanics, edited by Bryce S. DeWitt and Neill Graham (1973). Princeton University Press.

20. David Deutsch (1997). The Fabric of Reality.

21. Science and Ultimate Reality: Quantum Theory, Cosmology, and Complexity (2004). Edited by John D. Barrow, Paul C. W. Davies and Charles L. Harper, Jr. Cambridge University Press.

22. J. A. Barrett (2001). The Quantum Mechanics of Minds and Worlds. Oxford University Press.

23. S. Saunders, J. Barrett, A. Kent and D. Wallace (ed.) (2010). Many Worlds? Everett, Quantum Theory, and Reality. Oxford University Press.

24. Merriam-Webster.com Retrieved June 21st, 2013, http://www.merriam-webster.com/dictionary/epigenetic

25. Cai, et al. (2006). Postreactivation Glucocorticoids impair recall of established fear memory. Journal of Neuroscience. 26(37):9560-9566.

26. Gould E., Tanapat P., McEwen B.S., Flugge G., Fuchs E. (1998). "Proliferation of granule cell precursors in the dentate gyrus of adult monkeys is diminished by stress."

27. Stancampiano R., Cocco S., Cugusi C., Sarais L., Fadda F. "Serotonin and acetylcholine release response in the rat

hippocampus during a spatial memory task." Neuroscience. 1999;89(4):1135-43. Department of Biochemistry and Human Physiology, University of Cagliari, Italy.

28. Jacobs,T.L., et al., Intensive meditation training, immune cell telomerase activity, and psychological mediators. Psychoneuroendocrinology (2010)

29. Stuart Wilde. The Force.

30. Alan Watts. You're It: On Hiding, Seeking and Being Found.

31. Judy Willis (2009). How to Teach Students About the Brain, Educational Leadership, 67(4).

32. Nature 477, 23-25 (2011) doi:10.1038/477023a

33. Xiao-song Ma, et.al. Experimental delayed-choice entanglement swapping. Nature Physics 8, 479–484 (2012) doi:10.1038/nphys2294

34. File:Two-Slit Experiment Particles.svg, http://commons.wikimedia.org/wiki/File:Two-Slit_Experiment_Particles.svg, By inductiveload (Own work (Own drawing)) [Public domain], via Wikimedia Commons

35. Champagne, F.A.; Curley, J.P. (2005). "How social experiences influence the brain". Current opinion in neurobiology 15 (6): 704–9. doi:10.1016/j.conb.2005.10.001. PMID 16260130].

36. Skyrms, B., (1976) 'Possible Worlds, Physics and Metaphysics', Philosophical Studies 30, 323-332.

37. Hartle, J. B., (1968) 'Quantum Mechanics of Individual Systems', American Journal of Physics 36, 704-712.

38. Graham, N., (1973) 'The Measurement of Relative Frequency', in De Witt and N. Graham (eds.) The Many-Words Interpretation of Quantum Mechanics, Princeton NJ: Princeton University Press.
39. Gell-Mann, M., and Hartle, J. B., (1990) 'Quantum Mechanics in the Light of Quantum Cosmology', in W. H. Zurek (ed.), Complexity, Entropy and the Physics of Information, Reading: Addison-Wesley, pp. 425-459.
40. Gleason, A. M., (1957) 'Measures on the Closed Subspaces of Hilbert Space', Journal of Mathematics and Mechanics 6, 885-894.
41. Albert, D., (1992) Quantum Mechanics and Experience, Cambridge, MA: Harvard University Press.
42. Albert, D., and Loewer, B. (1988) 'Interpreting the Many Worlds Interpretation', Synthese 77, 195-213.
43. O'Connell, A. D. et al. Nature doi:10.1038/nature08967 (2010).
44. DiCarlo, L. et al. Nature Advance online publication doi:10.1038/nature08121 (2009).
45. Wan, L., Friedman, B. H., Boutrous, N. N., Crawford, H. J. (2008). P50 Sensory Gating and Attentional Performance. International Journal of Psychophysiology, 67: 91-100.